THE UPSTARTS

ALSO BY BRAD STONE

The Everything Store
Gearheads

THE UPSTARTS

How Uber, Airbnb, and the Killer Companies of the New Silicon Valley Are Changing the World

BRAD STONE

Little, Brown and Company
New York • Boston • London

Little, Brown and Company
Hachette Book Group
1290 Avenue of the Americas, New York, NY 10104
littlebrown.com

First Edition: January 2017

Little, Brown and Company is a division of Hachette Book Group, Inc. The Little, Brown name and logo are trademarks of Hachette Book Group, Inc.

The publisher is not responsible for websites (or their content) that are not owned by the publisher.

The Hachette Speakers Bureau provides a wide range of authors for speaking events. To find out more, go to hachettespeakersbureau.com or call (866) 376-6591.

ISBN 978-0-316-38839-9 (hc) / 978-0-316-39681-3 (large print) / 978-0-316-55456-5 (int'l ed.)
Library of Congress Control Number: 2016958932

10 9 8 7 6 5 4 3 2 1

LSC-C

Printed in the United States of America

For Tiffany

Upstart

up·start \'əp-ˌstärt\ (noun)

1. A newly successful person, business, etc.

2. A person who has recently begun an activity, become successful, etc., and who does not show proper respect for older and more experienced people or for the established way of doing things.

—Adapted from *Merriam-Webster's Learner's Dictionary*

Contents

THE UPSTARTS

Introduction

It was the beginning of something remarkable. Nearly two million people poured into Washington, DC, the week of January 19, 2009, for the inauguration of President Barack Hussein Obama. But not everyone there was just bearing witness. Among the throngs that gathered to brave the mid-Atlantic-winter chill, two groups of young entrepreneurs from San Francisco were on the verge of not just watching history but making it.

The three founders of a little-known website called Airbedandbreakfast.com decided to attend at the last minute. Brian Chesky, Joe Gebbia, and Nathan Blecharczyk convinced a friend, Michael Seibel, the CEO of the streaming-video site Justin.tv, to go with them. They were all in their midtwenties and had no tickets to the festivities, or winter clothes, or even a firm grasp of the week's schedule. But they thought they saw an opportunity. Their company had limped along for over a year with little to show for it. Now, the eyes of the world would be on the nation's capital and they wanted to take advantage.

They found a cheap crash pad in DC, an apartment in a drafty three-floor house near Howard University that, like so many other homes during that desperate time, was in foreclosure. The rooms were unfurnished save for a pullout sofa, which the three founders gave to Seibel. At night they crowded onto the hardwood floor on airbeds (naturally) along with their host, the manager of a local restaurant.

Their host was actually a tenant waiting for his inevitable eviction. He lived in the basement apartment and had used the AirBed & Breakfast website to rent out the empty first floor and, to three other guests, his own bedroom, living room, and walk-in closet. Sensing a promotional opportunity, Chesky e-mailed the staff of *Good Morning America* about the closet, and a producer promptly included it in a roundup of unusual accommodations for the inauguration.[1]

By day the founders and Seibel passed out AirBed & Breakfast flyers at the Dupont Circle Metro station. "Rent your room! Rent your room!" they cried to the bundled-up commuters, who mostly ignored them. At night they met other AirBed & Breakfast hosts in the city, attended any inaugural parties they could get into, and answered multiple e-mails from a disgruntled customer—the guest in the basement bedroom. The woman had driven her Volkswagen bus from Arizona to DC with her support dog, a Chihuahua, and she apparently wasn't too keen on the crowded accommodations. In a barrage of messages to the company's e-mail account that week, she complained that she was certain she smelled marijuana, that the juice she'd left in the fridge had been taken, and that the house didn't comply with Americans with Disabilities Act regulations.

At one point she threatened to call the police. The founders of the company sat just a few feet above her head, trying as best they could to assuage the anger of one of their few actual customers.

On the day of the inauguration, the group awoke at 3:00 a.m. to try to claim a good viewing spot in the National Mall. They walked two miles to get there, buying warmer coats, hats, and face masks at a kiosk in front of a Metro stop along the way. By 4:00 a.m. they had found a space on the green in the area open to the general public, a few football fields away from the presidential podium.

"We just kind of sat back to back in the middle of the Mall and tried to stay warm," recalls Brian Chesky, now the billionaire CEO of that once-fledgling company, Airbnb. "It was the coldest morning of my life. Everyone cheered when the sun came up."

* * *

Garrett Camp and Travis Kalanick also attended the festivities that week, and their experience was nearly as ignominious. A friend on the inaugural committee, the investor Chris Sacca, had convinced them to come. Kalanick, a Los Angeles native who had recently sold his startup to the web infrastructure company Akamai, made a twenty-five-thousand-dollar donation to the inaugural committee and split the expense with Camp. They were both in their early thirties, full of optimism about the coming transformative effects of technology despite the meltdown of the global economy. They were largely ambivalent about politics but didn't want to miss a historic moment or, just as urgently, a seminal party.

They were also unprepared for the trappings of a presidential inauguration. A few days before the event, they flew to New York City and went shopping for tuxedos at a Hugo Boss outlet. Wary of looking like twins, Kalanick went with a bow tie, Camp a regular tie.

The night before the inauguration, they ended up stranded in a line outside the Newseum, trying to get into a party hosted by the *Huffington Post*. It was windy and cold and they had only one wool hat between them, which they took turns wearing, ten minutes each, back and forth, while frantically texting one of the party's hosts, asking to be allowed inside.

On the big day, instead of waking up early like the Airbnb founders, Camp and Kalanick woke up late. Kalanick had rented a swank home near Logan Circle on the vacation-rentals website VRBO but it was a few miles away from the Mall, and no taxis were readily available. They ended up sprinting down the wide DC avenues for thirty minutes, side by side. When they finally got to their seats, perched with Sacca and his high-powered Silicon Valley friends above the inaugural platform, the sweat on their bodies cooled, giving way to an unbearable chill.

"By the end of the day, I was definitely sort of pre-hypothermic," Kalanick recalls. "Everyone was like, 'What's wrong with you?' and

I'm like, 'I'm frozen.'" Camp adds, "I grew up in Canada. I've been very cold. But that was one of the coldest experiences in my entire life."

At the time, Camp had been trying to get Kalanick excited about a business idea he was developing that would allow anyone with a smartphone to call a black town car with a click of a button. Kalanick had been interested but not particularly enthusiastic, conceding that it was a good idea, just not necessarily a big one. Yet here was concrete evidence that such a service was needed. A car that could be summoned on demand from a phone, Camp noted, could be vital in big cities when other transportation options were unavailable.

"See?" Camp said to Kalanick as the crowd chanted "O-bam-a! O-bam-a!" and the world waited for the new First Family to take the stage. "We really need this."

Even back then, Camp was calling this proposed service by a name the world would soon know well: Uber.

That was eight years ago.

Much has changed since then—the president, for starters. But few changes have been as profound as those that were ushered in by those two groups of entrepreneurs sitting anonymously in the crowd that day.

They had plenty of help. The late Apple co-founder Steve Jobs introduced the first iPhone seven months before Obama's inauguration. Two months after it, Jobs announced that the iPhone would run software programs, called mobile applications, or apps, from other companies. Other significant technology trends were converging at the same time. The social network Facebook, founded in a Harvard dorm room in 2004, was skyrocketing in popularity and persuading internet users to establish their identities online. The search giant Google was making it easier for other companies to integrate its mapping tool, Google Maps, into their own apps and websites. Computers and phones were getting cheaper and more powerful. Broadband internet use was skyrocketing.

All these intersecting trends produced the biggest tectonic shift in computing history since the invention of the web browser. In the span of ten years, a majority of the people in the modern world started to run large portions of their lives online, mostly through the slender slabs of plastic, glass, and silicon that they could hold in their hands and slip into their pockets.

The juggernauts Uber and Airbnb did not generate this technological wave, but more than any other companies over those eight years, they rode it and profited from it. The two companies, both in San Francisco, their headquarters only a mile apart, are among the fastest-growing startups in history by sales, overall market value, and number of employees. Together they have scrawled in the annals of entrepreneurship the most memorable stories of a third phase of internet history—the post-Google, post-Facebook era of innovation that allowed the digital realm to expand into the physical one.

They have attained these heights despite the fact that their businesses own little in the way of physical assets. Airbnb can be considered the biggest hotel company on the planet, yet it possesses no actual hotel rooms. Uber is among the world's largest car services, yet it doesn't employ any professional drivers or own any vehicles (save for a small, experimental fleet of self-driving cars). They are the ultimate twenty-first-century internet businesses, bringing not only new opportunities but new kinds of risks, often poorly understood, to those who provide and utilize their services.

Uber, as the world well knows, allows anyone to summon a vehicle with ease, track its progress on a virtual map, and then ride with a driver whose reliability is illustrated by a one- to five-star rating. The rider pays without the awkward exchange of cash or the time-consuming swipe of a credit card. The brilliance of this seamless transaction is so widely accepted in the LCD-lit halls of Silicon Valley that it inspired a surge of similar businesses in the fields of food delivery, package pickup, babysitting services, and so on.

Airbnb has extended the experience of traveling abroad well beyond the manicured realm of the hotel and the central tourism district. The concept was simple: Allow anyone to make a spare couch, unused bedroom, vacant in-law apartment, or empty second home available to a traveler on a short-term basis. This idea was not necessarily novel (VRBO, HomeAway, Couchsurfing, and Craigslist did it first), but the elegance of the solution was unrivaled. Carefully selected photographs and reviews from previous transactions introduce the host and the guest to each other before they meet in person. Like Uber, cash is deleted from the equation; Airbnb collects the transaction fee from the guest when the lodging is booked and remits it to the host, minus its cut, after the stay is complete.

During those eight years, the two companies etched their brands into popular culture. Their names are nouns and occasionally verbs, used by retirees looking to earn extra money, millennials seeking authentic travel experiences, and young people who have no interest in owning expensive assets like cars. Uber has become a staple of rap songs (Drake: "'Bout to call your ass a Uber, I got somewhere to be") and late-night monologues (Jimmy Kimmel: "About a quarter of Uber drivers are over fifty and many are much older than that. I guess you could think of it like Miss Daisy driving you").

Airbnb has drawn plaudits from President Obama himself. "I just want to brag on Brian just for one second," he said at a press event with Chesky in Cuba on March 21, 2016, during the first trip to Cuba by an American president in over eighty years. "He's one of our outstanding young entrepreneurs who had an idea and acted on it."

The companies' stories are different in many ways but similar in a few crucial ones. Their founders' original motives were not stated in high-minded terms, like Google's ("Organize the world's information and make it universally accessible and useful") or Facebook's ("Make the world more open and connected"). Camp, Kalanick, and their friends wanted to ride around San Francisco in style. Chesky and his cohorts were looking for a way to make some extra cash when a conference came to town.

Both startups offered age-old ideas (share a vehicle, rent your home) with new twists and ended up fostering a remarkable degree of openness among people who had never previously met. In a previous decade, most of us would have stayed far away from someone's private car or unlit home, scared by headlines about crime and by our mothers' earnest warnings to avoid strangers. Airbnb and Uber didn't spawn "the sharing economy," "the on-demand economy," or "the one-tap economy" (those labels never quite seemed to fit) so much as usher in a new trust economy, helping regular folks to negotiate transportation and accommodations in the age of ubiquitous internet access.

The nearly simultaneous emergence of both companies has been striking. For most of its first year, Airbnb was a side project that many dismissed as wildly outlandish. Why would a sane person ever want to sleep in a stranger's bed? Eight years later, investors valued the company at thirty billion dollars, more than any hotel chain in the world. Those founders who slept on the hardwood floor in Washington, DC? They are worth about three billion dollars each, at least on paper.[2]

Uber's potential was underestimated even by its own creators, who saw the service as a useful tool in San Francisco, a city whose cab industry badly underserved the needs of a booming business capital. But the startup exploded out of San Francisco and into New York, Los Angeles, Chicago, London, Paris, Beijing, and nearly every other major city. Early adopters raved about it to friends, who then signed up. As the company introduced less expensive varieties of the service, substituting regular cars for town cars and carpools for individual rides, many people came to rely on it. Uber's valuation in late 2016 was sixty-eight billion dollars, more than any other privately held startup company in the world. Kalanick and Camp have an estimated net worth of more than six billion dollars each.

Both companies' journeys have been marked by nearly nonstop controversy. In many cities, Uber sidestepped laws requiring professional drivers to undergo rigorous training sessions, submit to fingerprint-based background checks, and acquire expensive, government-issued chauffeur licenses. It has encountered fierce resistance from taxi companies and lawmakers and been the focus of violent protests. Cabbies have shut down the Autobahn in Berlin, blockaded the roads around Orly Airport in Paris, beaten up Uber drivers in Milan, and menaced Uber employees in Mumbai. There are fresh battles every month, exacerbated sometimes by the startup's own flat-footed moves and relentless grow-at-all-costs mentality and other times by the fiery resentment of incumbent taxi companies that have watched their businesses change with breathtaking speed. Uber is also the target of hundreds of lawsuits, many of which concern the legal status of its drivers, who are designated by the company as contractors, not employees. They get to set their own hours but enjoy none of the security of permanent employment.

Airbnb's rise has been just as eventful. The company encountered laws in, among other cities, New York, Barcelona, Amsterdam, and Tokyo designed to thwart illegal hoteliers and limit the number of nights per year that people could rent out their homes. Lawmakers, activists, and hotel unions have criticized the company for worsening the housing shortage in desirable urban areas, driving up housing costs, and skirting hotel taxes. In late 2016, Airbnb actually sued New York City and its hometown, San Francisco, over legislation that threatened the company and its hosts with thousand-dollar fines every time a prospective host posted a listing on the service that violated the cities' short-term-rental laws.

Together, these companies have come to embody a new business code that has forced local governments to question their faithfulness to the regulatory regimes of the past. Taxi medallions, municipal licenses to operate a cab, were an early twentieth-century invention meant to prevent an excess of cars on clogged city streets and assure riders that drivers were trained and vetted and knew how to navigate the city. Zoning laws and hotel and guesthouse ordinances kept commercial activity out of residential neighborhoods and ensured that hotel rooms were up to safety code. Airbnb and Uber substituted the self-policing tools pioneered by internet marketplaces like eBay—riders graded their drivers and guests evaluated their hosts, and vice versa.

Uber and Airbnb have also come to represent, at least to some, the overweening hubris of the techno-elite. Critics blame them for everything from destroying the basic rules of employment, exacerbating traffic, and ruining peaceful neighborhoods to bringing unrestrained capitalism into liberal cities. Some of that is overheated, but there were consequences to their approaches that even Uber and Airbnb did not anticipate.

At the center of this maelstrom are the young, wealthy, charismatic chief executives: Travis Kalanick and Brian Chesky. They represent a new kind of technology CEO, nothing at all like Bill Gates, Larry Page, and Mark Zuckerberg, the awkward, introverted innovators who typified the previous generation of tech leaders. Instead, they are extroverted storytellers, capable of positioning their companies in the context of dramatic progress for humanity and recruiting not only armies of engineers but drivers, hosts, lobbyists, and lawmakers to their cause.

Both were relatively unknown before moving their startups into the global business vanguard. Yet both have demonstrated extraordinary levels of ambition and boldness and a willingness to bet big despite the prospect of humiliating failure.

So how did it all happen? How did they maneuver past entrenched, politically savvy incumbents to succeed where others had failed and build large companies in a staggeringly short amount of time? How much of their success was luck? What does it take to survive and thrive in modern Silicon Valley?

These were questions, I judged back in 2014, that were ripe for exploration in a book. But the practical question was this: Would the startups cooperate with such an in-depth project? At most Silicon Valley tech companies, the time and public images of the top executives are obsessively guarded. Uber and Airbnb had graduated into this secretive sanctum.

The only way to find out was to ask.

* * *

Living up to its mission to foster hospitality, Airbnb promptly invited me to discuss the project. I met Brian Chesky at his company headquarters at 888 Brannon Street in San Francisco, a lavishly renovated former battery factory. The entrance to the building is majestic, with a five-story open atrium that is both striking and impractical; one three-story stretch of wall is furnished with a variety of plants that need almost constant care. Airbnb occupies several floors that have inspirational sayings etched on the walls and conference rooms decorated to look like exotic home rentals on the site.

I met Chesky in the Founders' Den, a wood-paneled holdover from a previous tenant, a paper-distribution company. Four brown leather armchairs surrounded a circular coffee table that sat on an oriental rug. It was an anachronism from the 1950s amid the splendor and excess of San Francisco's twenty-first-century internet boom. Across the street, cranes were assembling new high-priced condominiums.

Chesky is five foot nine and fit from a regular workout regime. He spoke quickly, with spasms of tension occasionally passing over his mouth, and related the history of his company's meteoric rise in terms of the moments of its most dramatic adversity.

Early in Airbnb's history, he said, "It felt like the world was against us and everyone was laughing at us." The startup persevered through widespread rejection by investors, battled a ruthless European rival, and survived a deluge of negative publicity around the early destruction of one of its hosts' homes by an unruly guest. "No one believed in us. We were insecure and had no idea what we were doing," he told me.

More recently, the primary adversaries have become regulators and housing activists. Some of them are looking to score political points by vilifying a high-profile target; others are legitimately worried about Airbnb's impact on housing affordability. Unlike his friend Travis Kalanick, Chesky presents himself as a sympathetic ally of those in the latter category. "We want to enrich cities. We don't want to be an enemy of affordable housing," he said. "I think we can be on the right side of the argument. We let many of our users stay in their own homes. This is why we were founded. If I didn't need the money to pay rent, we wouldn't have started the company."

As for the book project, he was game. Over the ensuing year, I spoke to Chesky, his co-founders, and top Airbnb executives. The company's PR representatives were helpful though understandably nervous about the outcome, soliciting questions for interviews ahead of time, sitting in on conversations, and taking copious notes.

Then there was the monumental challenge of obtaining the cooperation of the famously combative Travis Kalanick, known as a contrarian who advocated fiercely for his company's interests. He did not disappoint. "I came to this meeting out of respect for you and your work," he said when we met for dinner in March 2015 at the Burritt Room and Tavern in San Francisco's Mystic Hotel. "But I'm going into it thinking, *There's no way in hell I'm cooperating with a book about Uber right now*."

Kalanick had endured a year of negative press over Uber's tactics toward rivals, its ambiguous impact on cities, and its tense relationship with drivers. David Plouffe, Obama's former campaign manager and at the time Kalanick's chief of media relations, came along for the dinner and wore the bemused smile of someone witnessing a journalist's suicide mission.

Despite the inauspicious start, Kalanick seemed willing to listen. He asked what he had to gain by cooperating. "If you want people to embrace a radical future in which they give up their cars," I argued, "you have to allow journalists to explain and demystify your story. If you want to change the way cities work, Uber must be understood."

It didn't work. "You have to inspire me!" he said. "Tell me what we have to gain!" He was forthright and transactional; in other words, Travis being Travis.

At some point during the rye whiskey cocktails and flat-iron steaks with garlic-paprika fries, he seemed to warm briefly to the cinematic potential. "You'd start this story with a city council meeting," he mused. "The city council people are sitting up in front of the room and they are misinformed. They are thinking mostly about where their next campaign contribution comes from. There's an Uber rep there, but he is basically alone, trying to describe an unfamiliar and strange technology to people who have no understanding of it.

"The Uber guy has a lobbyist, but the lobbyist is also working for the other guys on the side. Finally, you have the big taxi guys there, and they have the city council locked up and paid for.

"Then, meanwhile, you'd cut to the taxi guys at the airport. They are all waiting there for hours, playing cards or whatever, for the chance to pick up one fare. And there's an Uber recruiter there, and he is surrounded, explaining this new system to the drivers..." Kalanick caught himself and trailed off. "Anyway, that's how you'd start the movie."

On the street after dinner, he said again, "You have to inspire me," as if I hadn't just spent two hours laying out my best arguments. Then he and Plouffe took off on foot, back to the office.

Six months passed, and despite my repeated entreaties, I didn't hear anything. But then, after I'd talked to dozens of regulators, competitors, and current and former Uber employees, a new Uber PR executive somehow convinced Kalanick to cooperate. I eventually spoke with two dozen different Uber executives from all periods of the company's brief history and had another few hours of Kalanick's time to complement several interviews I had conducted with him over the course of my five years as a writer for *Bloomberg Businessweek*.

The result is this book. It is not a comprehensive account of either company, since their extraordinary stories are still unfolding. It is instead a book about a pivotal moment in the century-long emergence of a technological society. It's about a crucial era during which old regimes fell, new leaders emerged, new social contracts were forged between strangers, the topography of cities changed, and the upstarts roamed the earth.

PART I

SIDE PROJECTS

CHAPTER 1

THE TROUGH OF SORROW

The Early Years of Airbnb

> Every great startup starts as a side project that isn't anybody's main priority. AirBed & Breakfast was a way to pay our rent. It was a way to pay rent and buy us time and help us get to the big idea.
>
> —Brian Chesky

The first guest to use Airbedandbreakfast.com was Amol Surve, a recent graduate of the Biodesign Institute at Arizona State University.[1] He arrived at his rental late in the afternoon on Tuesday, October 16, 2007, fifteen months before Barack Obama's historic inauguration, and was greeted at the door by the site's twenty-six-year-old co-creator Joe Gebbia, who politely asked him to remove his shoes.

Gebbia gave him a tour of unit C on the top floor of 19 Rausch Street, a narrow, side-street row house in San Francisco's chaotic South of Market neighborhood. It was spacious, with three bedrooms, two bathrooms, a comfortable living room, and, up the main flight of stairs, a roof terrace overlooking the golden city, which was undergoing its own momentous reinvention. At the time, the two men had no way of knowing that over the next few years, this apartment would be ground zero for a worldwide social movement and global business phenomenon called the sharing economy.

Surve, a native of Mumbai, India, had used the internet to rent an airbed for eighty dollars a night during the World Design Congress, a biennial conference held by the International Council of Societies of Industrial Design, or ICSID. All the hotels in the city that week were either booked or too expensive for Surve, so he hadn't expected much. But what he saw in his temporary home was promising. There was a shelf full of design books and a comfortable couch in the living room. He was invited to help himself to cereal and milk in the kitchen in the morning, and there was a small bedroom with an inflated airbed, sheets, and blankets. His hosts were surprisingly thoughtful; Gebbia presented him with a small bag that contained, among other things, the house rules, a Wi-Fi passcode, a city map, and some loose change for the neighborhood homeless population.

But by far the most surprising thing Surve saw that first afternoon was an image on Gebbia's laptop—a picture of Amol Surve himself. Gebbia and his roommate and business partner, Brian Chesky, were putting together a presentation about their new home-sharing service for a pecha-kucha (Japanese for "chatter"), an event in which a series of designers present their new product ideas by showing twenty slides apiece and discussing each slide for twenty seconds.

As the very first guest of this new service, Surve had been included in the presentation. His stay hadn't even begun yet and here he was, grafted into chapter 1 of what his hosts were clearly hoping would be a very long story. "It was very strange," Surve says years later.

Surve was happy just to have a comfortable place to sleep, but he ended up getting an education in the Silicon Valley startup scene as well. He spent lots of time that week on the couch with Gebbia and Chesky, talking about design and examining Apple's new device, the very first iPhone. Surve hadn't even heard of Steve Jobs, let alone the iPhone, and he was totally unfamiliar with the litany of motivational Jobs sayings that Gebbia and Chesky frequently quoted, such as "We're here to put a dent in the universe."

With another guest staying at Rausch Street, Kat Jurick, Surve attended the pecha-kucha, and later in the week, Gebbia took him on a tour of the city, showing him sights like the famous winding block of Lombard Street and the farmers' market outside the Ferry Building. Gebbia, who liked to exhibit his design sensibilities with stylish items like colorful sneakers and trendy, oversize eyeglasses, sported an aviator hat with furry ears in the fall chill.

After the conference, Surve had an extra day in the city and wanted to see the famous d.school—the Institute of Design at Stanford University. Chesky wanted to see it too, so he offered to drive. At Stanford, the pair sat in the front row at a free lecture given by the Italian designer Ezio Manzini and afterward introduced themselves to Bill Moggridge, a co-founder of Ideo, the iconic design firm, and chairman of that week's design conference.

It must have been an unusual sight: Moggridge, who passed away in 2012, was over six and a half feet tall. Chesky, who had the beefy frame of a hockey player and was a workout fanatic, was nine inches shorter. Staring up, the fast-talking Chesky launched into a description of AirBed & Breakfast, suggesting it could be the official accommodation for the Industrial Designers Society of America. As part of his impromptu pitch, he introduced Surve as the first guest, and Surve was once again conscripted into the story. Surve remembers Moggridge nodding without comment and looking skeptical.

Chesky would later say that AirBed & Breakfast was a lark and something of a side project at the time, but Surve remembers his new friend vibrating with enthusiasm for the idea on the forty-five-minute drive back to the city. "Amol," Chesky told him in the car, "we have to put a dent in the universe with this concept."

Brian Chesky grew up in an eastern suburb of Schenectady, New York: Niskayuna, a town no one had heard of located outside a city most people couldn't find on a map. His family was solidly middle class and lived in a five-bedroom Colonial house with a dog and a large backyard. His mom and dad, descendants of Italian and Polish immigrants, respectively, were social workers who doted endlessly on Chesky and his younger sister, Allison. When they weren't at their jobs (where they occasionally bent protocol by inviting the individuals and families they counseled into their home), they spent their time catering to their children. "We had no life," says Brian's vivacious mother, Deb. Adds his father, Bob, "Some people invest in their career. We invested in our kids."

From a young age, Chesky gravitated toward drawing, paying frequent visits to the Norman Rockwell Museum, about an hour's drive from his town. His parents marveled at his ability to sit and draw for long periods, and teachers favorably compared his style to Rockwell's and made heady predictions about his future. "Your son is going to be famous one day," one told them.

He also played hockey and imagined himself as the next Wayne Gretzky. He was quick and agile and drew recognition around the region. But after he broke his collarbone twice, his high-school coaches finally decided that he was too short and not strong enough to have a real future on the ice. His parents seemed to agree. "He was too small to be a star," says Deb.

Unwilling to accept defeat, Chesky started working out and lifting weights, chugging creatine-and-egg-white shakes and adding muscle.

During his college years, he joined the bodybuilding-contest circuit, oiling up and flexing onstage in front of cameras and crowds at national bodybuilding competitions. "I did this before I realized the consequences of the internet," he said later, sheepish about the photographic evidence from that period of his life.[2]

Chesky's friend and co-founder Joe Gebbia was born in Atlanta, Georgia, the youngest child of two self-employed sales reps who worked with independent health-food supermarkets across the south. With his older sister he would accompany his father on long road trips into Alabama, Tennessee, and South Carolina, peddling fruit and organic juices and sticking around after meetings to help store owners restock their shelves. Like Chesky, Gebbia straddled different worlds growing up. He played tennis and basketball and ran track and field but also studied the violin until deciding that all he wanted to do was play jazz piano like his idol Dave Brubeck.

Then one summer in the middle of high school, Gebbia took art classes at Valdosta State University in Georgia and decided that he really wanted to be a painter. "You've got something there," noted an instructor who admired his work, and he suggested Gebbia apply to one of the top art schools in the country, the Rhode Island School of Design. Gebbia spent the following summer taking classes at RISD and was enthralled by the majestic French and Neocolonial buildings clustered on the banks of the Providence River. Gebbia enrolled at RISD in 2000, a year after Chesky.

They met in classes and at student events and found each other easy collaborators. Chesky ran the school hockey team, the Nads ("Go, Nads!"), while Gebbia managed the basketball team, the Balls ("When the heat is on, the Balls stick together"). Running a RISD sports club was more a marketing challenge than a competitive one; the way they tell it, the teams cared less about winning than about using the games as excuses for campus high jinks.

Both gravitated toward the study of industrial design and the idea that they could make things that were as iconic and affordable as the classic Eames lounge chair. "You can live in a world of your own design," their professors insisted. "You can change the world, you can redesign it."[3] The department pounded a kind of practical idealism into the students' heads; during one field trip, they were taken in buses to the city dump and driven through the caverns of trash so they could see where wasted effort ended up.

Chesky and Gebbia teamed up one summer to work on a project for the hair-dryer maker Conair and on another idea that bodybuilder Chesky dubbed the Chesky Solution. He had a notion to use PalmPilots and other mobile gadgets, along with body sensors, to track people's health. Neither project went anywhere, but the duo solidified their friendship during long creative brainstorming sessions. "Everything for me came together because we had such a fun time working on that project," says Gebbia, who was always on the lookout for a business partner. "Our ideas were so original, and different than everybody else's."

Chesky was selected by his classmates to give the commencement address in 2004. In a video of the speech, he can be seen storming onto the stage to Michael Jackson's "Billie Jean." Full of charisma and confidence, he strips off his graduation robe to reveal a white sport coat and matching tie, does a few dance moves, and flexes his muscles. Then he cracks up the audience for the next twelve minutes. "Parents, I want you to know that investing in us is better than any stock on the market," he said presciently. "Sure, you spent a hundred and forty thousand dollars so we could paint with Jell-O and roll around in Silly Putty. But more important, you knew we needed to be inspired and we found plenty of it here at RISD."

Before Chesky left town to go back to Niskayuna, Gebbia took him for pizza and shared a prediction: One day, they would start a company together and someone would write a book about it. "I saw this gift he has, to be able to get people excited," Gebbia says. "I would have felt incomplete if I didn't express to him what I was feeling."

After graduation, Chesky spent a few months living at home, then he decamped to Los Angeles, moving in with former classmates in a Hollywood apartment amid the tourists and costumed panhandlers a few blocks from Grauman's Chinese Theatre. His parents, still doting, bought a Honda Civic from an L.A. dealer and had it delivered to him at the Los Angeles airport.

He was living his college dream. He had a real job, earning forty thousand dollars a year as a designer for the consultant shop 3DID in Marina del Rey, working on toys for Mattel, guitars for Henman, medical equipment, footwear, dog toys, and handbags.

"When you are a designer at school, especially an industrial designer, you just dream of getting something on the shelf," Chesky says.

But somehow that first job never quite matched his expectations. There was a ninety-minute commute each way in the suffocating gridlock on the I-405. Most of the projects he worked on never made it to stores or, if they did, ended up in landfills.

In 2006, Chesky's firm was invited to participate in the Simon Cowell–produced reality-TV show *American Inventor.* His team was tasked with aiding a husband and wife with their concept for a bacteria-free toilet seat, called the Pureflush. The couple was competing against a dozen other inventors for a million-dollar grand prize. Chesky and his colleagues from 3DID were there to help them conceptualize the product and produce a prototype.

The episode aired on May 4, 2006. Chesky's presence was largely edited out of the show, but you can see the future billionaire sitting quietly with colleagues in design reviews. In retrospect it is easy to see why the experience might have propelled Chesky farther down the path of youthful disillusionment. The husband, a part-time magician with a volcanic temper, started screaming when 3DID revealed its mock-ups of the toilet seat.

"This one is way too small. That one is obviously way too big! End of story! This bullshit stops right here!" he yells in the episode, to the shock of the designers (and, no doubt, the delight of the show's producers). "You are not here to make it better! You are here to take our dreams and put them in physical form!"[4]

For Chesky, that unhinged criticism was cutting. At the time he was obsessively following the story of the fantastically successful founders of the video-sharing site YouTube; he was spending hours on the site as well as watching Steve Jobs's keynote presentations and the television film *Pirates of Silicon Valley.* This was a universe where new things really did change reality. "I got kind of obsessed," he says. "I was living vicariously, escaping to a world where someone could build something and actually change something. I was not doing that. I was sitting in a dark office making stuff for closets and landfills."

By early 2007, Chesky was antsy. He had moved to a two-bedroom apartment in West Hollywood with four friends and cut his hours at the firm to focus on making his own furniture. He designed a fiberglass chair with curves inspired by the hood of a Corvette Stingray, as well as a portion-control plate, a dish with an elevated center that could hold only a moderate-size, three-ounce piece of meat.[5] Desperate to start something of his own, to make a mark and see his name in lights, he toyed with creating a design firm called Brian Chesky Inc.

But he couldn't shake the notion that it was all pointless and uninspired and that it wouldn't lead to the promised life that he'd found so seductive at RISD, or in the movie *Pirates of Silicon Valley,* or in the pages of the Walt Disney biography that he was avidly reading. "People told me, you can change the world you are living in," he says. "Then reality set in, and reality wasn't that. The reality was, I'm just making stuff."

Then one day in the summer of 2007, Chesky got a package from his old college friend Joe Gebbia, who was living in San Francisco. The package inspired him to escape his brewing malaise.

Gebbia had fared only slightly better than Chesky in his post-RISD career. During his freshman year at college he had hit upon what he thought was a novel idea. At RISD, during marathon reviews called Critiques, students often had to sit for hours on metal benches and hard wooden stools that were covered in charcoal dust and paint. When they uncomfortably lumbered to their feet at the end of these sessions, the seats of their pants were invariably stained. To address this scourge, Gebbia designed a colorful foam cushion with a handle and the imprint of a rear end. He called his invention CritBuns.

After graduation, Gebbia financed the manufacture of the cushions with the proceeds from a RISD design award and stored eight hundred units in the basement of his Providence apartment building. Then he naively set out to sell it to stores at $19.99 apiece. The first stop was the Brown University Bookstore. Gebbia wore his best suit and made an impassioned pitch to a store buyer, who let him talk for a minute, said "No, thanks," and walked away. Buyers at the second and third stores gave him the same response. "I met the face of rejection," Gebbia told me later. "You can say that rejection slapped me in the face."

Finally a boutique in downtown Providence agreed to stock four cushions. Gebbia ran home, brought them back, and that night pressed his face to the glass window and gazed at them admiringly.

CritBuns didn't fly off shelves or change the world (he says he only sold a few batches). But it did command a prime position in his portfolio and helped to score Gebbia a coveted internship at Chronicle Books, a large independent publisher in San Francisco, where he moved in 2006 to design books and gift packaging. After he settled into his adopted city, he sent his old friend Brian Chesky a package. It contained a CritBuns cushion.

To Chesky, the ridiculous foam seat represented something profound. Gebbia had started something—a real company with a real product! He had made an impact. That summer Chesky visited Gebbia for his birthday and found San Francisco enthralling. When Chesky woke up the next morning on the sofa, one of Gebbia's roommates, a tall, lanky programmer, was hunkered over his laptop, his hands flying over the keyboard writing computer code. People were actually making things here and trying to change the world.

That fall, that tall roommate moved out of the apartment on Rausch Street. Gebbia needed to replace the contribution to the monthly rent as soon as possible. He asked Chesky if he wanted to move in.

Chesky toyed with the idea of living in San Francisco over the weekends but keeping his life and a new part-time teaching job in Los Angeles. He asked Gebbia if he could rent the couch in the living room for five hundred dollars a month instead of paying more for the entire bedroom. Gebbia told him flatly that unless Chesky fully committed, he was going to have to give up the apartment.

Then one day in early September, Chesky woke with his mind made up. Walt Disney himself had taken a huge risk by moving from Kansas City to Hollywood in 1923, and his life had changed. Chesky would take a chance too.

Of course, by moving into Rausch Street, Chesky wasn't solving the quandary of how to pay the rent. He still didn't have meaningful employment, and both RISD graduates were, essentially, broke. So a few weeks later, on September 22, 2007, with the World Design Congress coming to San Francisco and the city's hotels either overbooked or overpriced, Gebbia sent Chesky the e-mail that would change their lives:

> Subject: subletter
>
> Brian,
>
> I thought of a way to make a few bucks — turning our place into a designer's bed and breakfast — offering young designers who come into town a place to crash during the 4 day event, complete with wireless internet, a small desk space, sleeping mat, and breakfast each morning. Ha! — Joe.

It took Chesky and Gebbia three days to put together the first Airbedandbreakfast.com website using the free tools on the blogging website WordPress. Their site was basic, displaying the service's name in blue and pink cursive font and with a brief description of a concept that had preciously few defined rules. "AirBed & Breakfast is an affordable housing resource, social networking tool, and an up-to-the-minute guide" to the conference, they wrote. "Designers can choose which designers to meet, stay with, and at what price. The terms are up to you!"

The founders e-mailed their site to the city's design blogs and got their first taste of publicity from writers who were bemused by the concept. "If you're heading out" to the design conference "and have yet to make accommodations, well, consider networking in your jam-jams," one wrote.[6]

Much later Chesky would weave a handy mythology around the stay of Amol Surve and two other guests at his Rausch Street apartment during the design conference. When the three travelers waved good-bye, his story went, the co-founders not only were able to pay the rent but were struck by the depth of the bonds they had forged with their guests and realized that their absurd idea was the seed of a much larger business. But that story, like all such startup mythologies, is not entirely accurate. The co-founders didn't need to pay rent as much as they needed, desperately, to find a business idea, validate their potential, and fulfill the promise of their RISD education. When Surve and the others left, Chesky and Gebbia returned to their daily lives and their attempts to find meaning in their mundane, post-RISD reality.

Part of that process involved meeting regularly with one of Gebbia's former roommates, that tall programmer who could type remarkably fast: Nathan Blecharczyk, a Harvard-educated engineer, only twenty-five but with a colorful entrepreneurial history that a few years later would prove valuable.

Blecharczyk had moved out of the Rausch Street apartment but had stayed in close touch with Gebbia. They had worked together on various projects, realizing that their programming and design skills complemented each other. Over the next few months the three young men met frequently and brainstormed ideas for new companies. One early notion was a roommate-matching service, combining elements of Facebook and the online-classifieds site Craigslist. A few weeks into developing the idea they discovered such a site already existed and was called Roommates.com.

Even though the three had been meeting often, Blecharczyk heard of AirBed & Breakfast only in January 2008, when Gebbia and Chesky visited his new apartment and asked him to join them. "You were really excited to tell me something but you wouldn't tell me what it was. You were being very secretive," Blecharczyk said to his co-founders during a joint interview years later. They went out for drinks and Chesky and Gebbia spilled their experience of the design conference and their idea—allowing people to share their homes during conferences and big events in major cities. They described a long list of features they wanted to build, including member profiles and ways for guests and hosts to rate each other. Blecharczyk, who was involved with a number of other projects at the time, was wary. It sounded like a lot of work, and as the only one of the three who had actual technical skills, he would be doing most of it. The pitch "left me with a fair amount of concern that this was not a realistic undertaking," he said.

A week later they met again at the Salt House, a new downtown restaurant, and Gebbia and Chesky presented a more modest version of their plan that could be completed by the upcoming South by Southwest Conference, a few weeks away.

Blecharczyk had downed a few drinks at dinner and impulsively agreed to build the site. But it still wasn't a priority for him.

Later that month he sent one of his semiregular group e-mails to friends and family, updating them on the projects he was working on. He listed a Facebook advertising network that he had conceived and another tool for the social network that allowed members to see which of their neighbors were using it. At the end of his e-mail, he noted as an afterthought that there were a few minor projects he was also working on, including a site called AirBedandBreakfast.com.

"I think it's a cool idea but probably not a big market," he wrote.

Chesky and Gebbia were among those who received the message. "We got that e-mail and we were like, 'What the fuck?'" Chesky recalls. Gebbia says it was "a punch in the stomach."

Nevertheless, Blecharczyk came through with a new version of a site on March 3, a week before the annual conference in Austin, Texas. The new slogan was "A friend, not a front desk."

There were of course no actual properties listed on the brand-new AirBed & Breakfast. So Chesky e-mailed anyone he could find in Austin who had listed rooms on Craigslist and invited them to post on the site. There ended up being two reservations for the conference—and his was one of them. He stayed with Tiendung Le, a Vietnamese PhD student who was studying construction engineering at the University of Texas and who lived in a two-bedroom apartment with his girlfriend in Austin's Riverside neighborhood.

Chesky stayed there for two nights on an airbed, which Tiendung Le and his girlfriend nicely furnished with a mint on the pillow. They also made him an espresso, which they remember he downed in a single gulp, and a bowl of Vietnamese noodles. But Tiendung Le, who now lives in Melbourne, remembers Chesky as distracted and jittery that week, often standing on the balcony and staring wistfully toward downtown, as if all the action were there and he was far from it. "He was not very much present in the place, in the sense that he seemed to be thinking about something else," Tiendung Le told me.

On the second morning of his stay, Chesky planned to hear Mark Zuckerberg speak at the conference, and Tiendung Le gave him a ride. On the way they talked about the young and successful Zuckerberg, who was rocketing to fame. Chesky bristled with excitement at the opportunity to hear him speak. (The talk, with blogger Sarah Lacy, was considered something of a famous bust when attendees started Tweeting angrily that the conversation lacked substance.) Along the way, Chesky thanked Tiendung Le for being "open-minded" and agreeing to try the apartment-sharing website. Tiendung Le was surprised by that and recalled it years later. "I was not aware of the fact that I was open-minded. We were students in Austin. We tended to be open to new things."

The next day Chesky left the apartment and decided to stay in Austin to meet one of Gebbia's former roommates, a man who worked for the video website Justin.tv and had a room at the

Hilton. Somehow, there was a miscommunication—Chesky couldn't find him, and late at night he ended up preparing to sleep in the hotel lobby.

But the friend and his colleague, a well-connected entrepreneur named Michael Seibel, finally found him and invited him up to their swank hotel suite. It was there, recovering after his brush with a night of inadvertent homelessness and somehow undeterred by his failure to drum up new business at the conference, that Chesky saw his luck finally start to change. It was late, and Chesky later recalled that Seibel was wearing only his underwear and that a TV program on the assassin John Wilkes Booth was playing in the background. But he started to pitch the AirBed & Breakfast concept again with renewed vigor. Seibel listened with curiosity and perhaps a bit of compassion. He would become the founders' first mentor, introducing them to investors and counseling them on how to draft their slide decks and polish their pitches.

"I know people who can write you a twenty-thousand-dollar check over dinner," Seibel boasted, and then he told him about angel investors, the class of financiers who got Silicon Valley tech companies off the ground. Chesky, still a tech-industry novice, would later say that he briefly thought Seibel was talking about actual angels.[7]

Chesky returned to San Francisco full of ideas for improving the website. He hadn't brought enough cash to Austin, and the awkwardness of paying Tiendung Le got him thinking about a way to introduce credit card transactions into the service.

But then, out of nowhere, Blecharczyk announced that he was moving back to Boston to live with his girlfriend, Elizabeth, a fourth-year student at Harvard Medical School. "I was excited about AirBed and Breakfast, but to me it was a side project, and one of a few," he says.

Between April and June of 2008, almost nothing happened with the fledgling business. Airbnb was very nearly stillborn. Then the idea dawned on Chesky and Gebbia that presidential candidate Barack Obama was set to talk to eighty thousand people at the Democratic nominating convention in Denver that August. The Mile-High City did not have nearly enough hotel rooms, and the eyes of the world would be on the convention.

From Boston, Blecharczyk recognized the unique opportunity and agreed to work on yet another version of the website between his other commitments. In the site's third iteration, the founders endeavored to make renting a room as easy as booking a hotel room. There was a search box that asked travelers where they were going, a large green book-it button, and sizable photographs of the hosts and their residences.

Every Friday that spring, Gebbia and Chesky brought mock-ups of the new design to Michael Seibel at Justin.tv. Seibel and his Justin.tv co-founder, Justin Kan, observed their progress, identified problems, and sent them away to make improvements (the early payment mechanism, they recalled, was a particular mess). Seibel and Kan weren't paid for this and received no equity in the fledgling startup. It was simply how things worked in Silicon Valley's cliquish network of founders. "On the East Coast you give money to charity," Seibel says. "On the West Coast in the startup world, if you want to give back, you help young founders. This is a game where karma matters."

By spending time at Justin.tv, the Airbnb founders got to see what a real tech startup looked like, one with real offices, real employees, and actual venture capital in the bank. (Justin.tv later spun off a video-game service, Twitch.tv, which was acquired by Amazon in 2014 for $970 million.) Continuing this education, they attended a one-day event called Startup School, organized by the startup incubator Y Combinator and hosted by Stanford University. The speakers that year included Amazon CEO Jeff Bezos and the investor Marc Andreessen, an inventor of the web browser. But the speech the founders remembered best was by Greg McAdoo, a venture capitalist at the top-tier VC firm Sequoia Capital, a man whom they would soon get to know well.

McAdoo spoke about why being a great entrepreneur required the precision of a great surfer.

> If you want to build a truly great company you have got to ride a really big wave. And you've got to be able to look at market waves and technology waves in a different way than other folks and see it happening sooner, know how to position yourself out there, prepare yourself, pick the right surfboard—in other words, bring the right management team in, build the right platform underneath you. Only then can you ride a truly great wave. At the end of the day, without that great wave, even if you are a great entrepreneur, you are not going to build a really great business.[8]

Early that summer Seibel finally came through on his boast and introduced the founders to seven angel investors. Chesky wrote to them, told them who he was, pitched the company, and asked for a hundred and fifty thousand dollars to bootstrap it. He received five outright rejections and later published these e-mails online.[9] Two investors didn't even bother to write him back. "Very few people even met with us," Chesky says. "They considered us crazy."

There were several in-person meetings, but those fared just as poorly. One investor, a former Google executive, met Chesky and Gebbia at a café in Palo Alto, ordered a smoothie, and began to listen to the pitch. Then he walked out in the middle of it, his drink practically untouched. Gebbia and Chesky were left sitting there, wondering if the investor would return.

In early August, Chesky and Gebbia were invited to the Palo Alto offices of Floodgate, the angel-investor firm that had backed Justin.tv. Even though only a dozen reservations a week were booked on the site, Chesky felt confident: AirBed & Breakfast had caught the attention of the influential industry blog *TechCrunch*.[10] Instead of giving a slide presentation, Chesky planned to demonstrate the site live. But when he stood up to talk, he realized with horror that traffic from the *TechCrunch* article had crashed the website. He ended up making small talk while Gebbia tried desperately to reach Blecharczyk, who already knew about the outage—the engineer had set up a service that sent him a text message with a single word—*AirbedDeflate*—every time the site went down. But it was too late. Chesky bombed, and Floodgate passed on the deal.

All these investors had concerns about the size of the market, about the absence of any real users, and about the founders themselves, who didn't resemble the wonky innovators who'd created great Silicon Valley companies, people like Mark Zuckerberg and Steve Jobs. Design students seemed risky; Stanford computer science dropouts were considered a much better bet. And, frankly, the idea itself seemed small. "We made the classic mistake that all investors make," wrote Fred Wilson, a Twitter backer, a few years later. "We focused too much on what they were doing at the time and not enough on what they could do, would do, and did do."[11]

The year 2008 was also an anxious time in Silicon Valley. The tech industry had recovered from the devastation of the dot-com bust a few years before and had been buoyed by the Google IPO in 2004 and the budding success of Facebook. Yet the global economy was teetering, with problems mounting in the real estate market and an economic collapse just months away. In October of that year, Sequoia distributed a presentation, commonly referred to as "R.I.P. Good Times," advising its startups to drastically cut back on spending, lower their risk, and reduce debt. Investors didn't believe in Airbnb, but more than that, they were wary in general.

Even when Airbnb seemed close to obtaining capital, things had a way of going awry. Paige Craig was a Los Angeles–based angel investor and former member of the Marine Corps who had been looking for opportunities in the hospitality market when he stumbled upon Airbnb that summer. He was impressed with the diligence and work ethic of the founders and was primed to make a $250,000 investment. They agreed on a valuation and even met for dinner in San Francisco to close the deal early that fall, but the next day, Chesky declined to sign the deal documents, and he later declined to say why. A person close to the discussions, however, said that at drinks after the dinner, Craig gave Chesky the impression that he would make a difficult partner.[12] In Silicon Valley it was orthodoxy that the right investor could empower a company but a difficult one would cause unending problems.

Years later, Paige Craig heard through another investor that the founders had concluded he was a "crazy marine" and gotten cold feet. "I'm not upset and understand where they were coming from," he e-mailed me when I asked him about the missed opportunity. "A Google search on me back then would have clearly painted me as 'dumb money.' It motivated me to work on building up my experience, building a founder-friendly brand and working my ass off to win future deals. But damn that was an expensive lesson."

For Chesky, the money must have been difficult to turn down. He says that at the time, he had never felt more like a failure. Among the co-founders, Blecharczyk had his personal projects, and Gebbia had CritBuns and work as a consultant. Chesky didn't have anything except his old furniture designs and the fervent belief that a great wave of connectivity and sharing was gathering momentum and that people were ready for this strange brand of internet-facilitated intimacy.

If Airbnb was going to take off, it needed to happen quickly. Chesky had spent his life savings and both he and Gebbia were getting deeper in debt. Certain they would raise money, they had been accumulating credit cards and blowing through the spending limits. Chesky kept his maxed-out credit cards in a shoe box, while Gebbia slid his into the plastic sleeves of a folder meant to hold baseball cards.

Their situation was precarious and Chesky knew it. "I woke up every morning with my heart pounding," he says. "I would convince myself over the course of the day that everything was going to be okay and I would go to bed feeling good. And I'd wake up every morning with my heart pounding, Groundhog Day, asking, 'How did I get into this situation? What did I do to myself?' "

The conventions that summer only temporarily allayed his anxiety. About eighty people used the service to stay in Denver and there were articles about it in *U.S. News and World Report*[13] and the *Chicago Sun-Times.*[14] About two hundred new hosts signed up every week in August, and Airbnb collected a commission of around twelve dollars for each hundred-dollar-per-night booking. But then, after the conventions, things quieted down, the number of new reservations booked each week dwindled below ten, and once again Chesky was waking up early, staring at the ceiling, and marinating in dread over his unfulfilled potential.

Silicon Valley's startup scientists have a name for this phase in a company's gestation; they call it the Trough of Sorrow, when the novelty of a new business idea wears off and the founders are left trying to jump-start an actual business. Gebbia and Chesky experienced a deep trough that would have swamped most founders. They responded in a characteristic way, digging back into their RISD past and tapping their penchant for reckless, silly creativity.

Talking over their dismal prospects one night in the Rausch Street kitchen during the presidential debates, they started riffing on the idea of making breakfast cereal and offering it to guests. It could be presidential-themed breakfast cereal! One could be called Obama O's: "The breakfast of change!" And the other Cap'n McCains: "A maverick in every bite!"

"This is probably where it should have ended," Gebbia says. But for some reason, mired in the Trough of Sorrow, they couldn't let it go. Gebbia called Kellogg's and General Mills; employees hung up when he excitedly described the concept. He called the local cereal distributors and got nowhere there either.

Eventually they just decided to produce it themselves. Gebbia found an alumnus of RISD across the bay in Berkeley who owned a printing shop, and he somehow convinced him to print a thousand boxes in exchange for a percentage of the sales. The boxes advertised themselves as limited editions and had playful games on the back along with information about AirBed & Breakfast.

Then Gebbia and Chesky went to a local supermarket in a low-income neighborhood and bought dozens of boxes of cereal (Honey Nut Cheerios for Obama O's and Fiber One honey squares for Cap'n McCains), maneuvering past the puzzled looks of cashiers. Back in their kitchen, they assembled the cereal boxes, scalding their hands with a hot-glue gun, and transferred in the sealed bags of cereal.

"As if it couldn't get any more ridiculous," Gebbia says, one day they got an e-mail from a host, a professional jingle maker, who offered to compose accompanying songs for their website. The jingles are still on YouTube:

> Well, there's a really cool cereal that you ought to know
> Everybody is talking about Obama O's,
> In just one bite you will understand,
> 'Cause every single O sings, Yes, we can![15]

Their collaborators did not react well to the cereal gambit. "Nate was in a state of disbelief" when they told him about it, Gebbia says. Michael Seibel was furious. "That was the first time I was really worried about them," he says.

Somehow, it paid off. Once again demonstrating a flair for showmanship, the founders mailed boxes to every media outlet they could think of, right at the height of the presidential news cycle. Sensing an eclectic story, reporters called them back. Orders for the cereal poured in and they sold out of Obama O's in three days.

The cereal operation allowed the founders to pay off the Berkeley printer and resolve most of their credit card debt. It did not propel the company to immediate success or generate any significant wealth; in fact, they were still barely making ends meet and began subsisting on the surplus Cap'n McCains. But it did demonstrate an extreme level of commitment and an ability to think creatively that, ultimately, would lead to their long-awaited break.

A few weeks later, Chesky decided that the founders of the struggling company should apply to the prestigious Y Combinator startup school, which invested seventeen thousand dollars in each startup, took a 7 percent ownership stake, and surrounded founders with mentors and technology luminaries during an intense three-month program. It was a last-ditch effort and Chesky actually missed the application deadline by a day. Michael Seibel, an alumnus of the program (and later its CEO), had to ask the organizers to let the company submit late. They got permission, and the co-founders were invited for an interview. Blecharczyk flew out to San Francisco and crashed on the living-room couch on Rausch Street, and the three co-founders gathered themselves for one last try.

"If we didn't get in, we would not exist," Gebbia says. "The business was just not working."

Before they left for the interview, Gebbia went to grab boxes of the cereal. Blecharczyk snapped at him. "No, no, no," he said. "Keep the cereal at home." Gebbia pretended to acquiesce, then surreptitiously slipped two boxes into his bag anyway.

The interview at Y Combinator's offices in Mountain View was practically hostile. "People are actually doing this?" asked Paul Graham, the program's legendary co-founder, when the three men described the home-sharing concept. "Why? What's wrong with them?" Graham, then forty-four, later admitted that he didn't get it. "I wouldn't want to stay on anyone else's sofa and I didn't want anyone to stay on mine," he says.

But after they turned to go, to Blecharczyk's consternation, Gebbia brought out the two boxes of cereal and handed them to Graham, who was rightfully confused. Then the whole tangled story of the last year came tumbling out, from the inspiration at the design conference to the disastrous South by Southwest to the nominating conventions and the unlikely cereal gambit. "Wow, you guys are like cockroaches," Graham finally said. "You just won't die."[16]

Cockroach was Graham's word for an unkillable startup that could weather any challenge, and it was the highest possible compliment in his startup lexicon. A few weeks later, after the founders had learned they had gotten into the program, and after they'd visited Washington, DC, for Barack Obama's historic inauguration, they arrived at the offices of Y Combinator. Graham was there, speaking with Greg McAdoo, the Sequoia venture capitalist who had delivered that memorable speech about great waves the previous year.

McAdoo and Graham were discussing that most essential characteristic of great entrepreneurs: mental toughness, the ability to overcome the hurdles and negativity that typically accompany something new. McAdoo and his partners had identified this kind of true grit as the most important attribute in the founders of their successful portfolio companies, like Google and PayPal.

Scouting for new opportunities despite the gathering economic storm enveloping the world, McAdoo asked Graham: "So, who in this class of startups is the most mentally, emotionally tough?"

"Well, that's easy," Paul Graham responded, and he pointed across the room at two designers and an engineer, all hunkered over their laptops. "Hands down, it's those guys over there."

CHAPTER 2

JAM SESSIONS

The Early Years of Uber

> When you open up that app and you get that experience of, like, I am living in the future, like I pushed a frickin' button and a car showed up and now I'm a pimp—Garrett is the guy who invented that shit! Like, I just want to clap and hug him at the same time.
>
> —Travis Kalanick[1]

The whole thing might not have happened without Bond—James Bond.

It was mid-2008, around the time Brian Chesky and Joe Gebbia were working their way through the early versions of AirBed & Breakfast. The Canadian entrepreneur Garrett Camp had just sold his first company, the website discovery engine StumbleUpon, to eBay for seventy-five million dollars. Now he was living large, enjoying San Francisco's nightlife, and when relaxing at home in his apartment in the city's tony South Park neighborhood, he occasionally popped in the DVD of Daniel Craig's first Bond movie, *Casino Royale.*

Camp loved the movie but something specific in it actually got him thinking. Thirty minutes into the film, Bond is driving his silver Ford Mondeo in the Bahamas on the trail of his nemesis, Le

Chiffre, when he glances down at his Sony Ericsson phone. It's brazen product placement and by today's standards the phone seems comically outdated. But at the time, what Bond saw on his phone startled Camp: a graphical icon of the Mondeo moving on a map toward his destination, the Ocean Club. The image stuck in his head, and to understand why, you need to know more about the restless, inventive mind of Garrett Camp.

Camp was born in Calgary, Canada. His mother was an interior designer and his father had left a career in accounting to train himself as an architect and contractor. The Camps were itinerant back in the 1980s; his father would build a house, his mother would decorate it, and then the family would move in for a few years before selling the house and starting over.

Camp spent his early childhood playing sports, learning the electric guitar, and asking lots of questions. The family didn't have a television until he was fourteen, but they did see movies. He remembers that after the first *Back to the Future,* he needled his father ceaselessly about how nuclear fusion worked.

Eventually his curiosity settled on the geeky world of personal computers. An uncle gave the family an early model Macintosh, from the days of floppy disks and point-and-click adventure games, and Camp spent hours with it during the frigid winters, toying with early computer graphics and writing basic programs.

By the time Camp graduated from high school, his parents had nearly perfected their craft with a three-story home that included a comfortable office and a computer room in the basement. "There wasn't much reason to leave," he says.

Camp enrolled in the nearby University of Calgary, saved money by living at home, and spent the next few years there (aside from one year in Montreal, interning at a company called Nortel Networks). He got his undergraduate degree in 2001 and stayed at the university to pursue a master of science, finally leaving his comfortable nest after he turned twenty-two to move into a campus apartment with classmates.

Camp met Geoff Smith, who would become his StumbleUpon co-founder, through one of his childhood friends and together they started the site as a way for users to share and find interesting things on the internet without having to search for them on Google. Camp was obsessed with collaborative information systems and the semantic web. He didn't go out much back then, splitting his time between his graduate thesis and the company and immersing himself in dense academic papers about esoteric topics in computer science.

By the time Camp finished his degree in 2005, StumbleUpon was starting to show promise. Camp and Smith met an angel investor that year who convinced them to move to San Francisco and raise capital. They incorporated the company in the United States, and over the next year, the number of users on StumbleUpon grew from five hundred thousand to two million.

With the trauma of the first dot-com bust fading and the scent of opportunity again wafting across Silicon Valley, acquisition offers for StumbleUpon started pouring in. In May 2007, eBay bought StumbleUpon for seventy-five million, turning it into one of the early successes of what became known as Web 2.0, the movement in which companies like Flickr and Facebook mined the social connections among internet users.[2] For Camp, it seemed at the time like the highest possible level of success in Silicon Valley, and it was, by any reasonable standard—until the one that he achieved next.

Camp continued to work at eBay after the sale, and he was now young, wealthy, and single, with a taste for getting out of the house more often. This is when he ran headlong into San Francisco's feeble taxi industry.

For decades, San Francisco had deliberately kept the number of taxi medallions capped at around fifteen hundred. Medallions in the city were relatively inexpensive and couldn't be resold, and owners could keep the permit as long as they liked if they logged a minimum number of hours on the road every year. So new permits usually became available only when drivers died, and anyone who applied for one had to wait years to receive it. Stories abounded about a driver waiting for three decades to get a medallion, only to die soon after.

The system guaranteed a healthy availability of passengers for the taxi companies even during slow times and ensured that full-time drivers could earn a living wage. But demand for cars greatly exceeded supply and so taxi service in San Francisco, famously, sucked. Trying to hail a cab in the outer neighborhoods near the ocean, or even downtown on a weekend night, was an exercise in futility. Getting a cab to take you to the airport was a stomach-churning gamble that could easily result in a missed flight. (Even when a passenger arranged for a taxi via phone, he couldn't be sure the cab would show up; the driver might decide to pick up a street hail instead.)

Attempts to improve the situation were fruitless, since the fleets and their drivers were adamant about limiting competition. Over the years, whenever the mayor or the city's board of supervisors tried to increase the number of medallions, angry drivers would fill city council chambers or surround city hall, causing havoc.

After the eBay acquisition, Garrett Camp splurged on a red Mercedes-Benz C-Class sports car, but the vehicle sat in the garage. He hadn't driven much in Calgary—his parents hadn't wanted to pay the extra auto insurance—and in college he preferred to take public transportation. "Driving in San Francisco was too stressful," he says. "I didn't want to park the car on the street and I didn't want people to break into it. Just logistically, it was much harder to drive."

So the city's sad taxi situation put a serious crimp in Camp's new lifestyle. Since he couldn't reliably hail a cab on the street, he began putting the yellow-cab dispatch numbers in his phone's speed-dial. Even that was frustrating. "I would call and they wouldn't show up and while I was waiting on the street, two or three other cabs would go by," he says. "Then I'd call them back and they wouldn't even remember that I called before. I remember being late for first or second dates. I could start getting ready twenty minutes early and still I'd end up being thirty minutes late."

The sparkling City by the Bay beckoned, but Camp had no reliable way to answer its call. Habitually restless, frustrated by inefficiencies, and armed with a willingness to challenge authority, Camp came up with his first attempt at a solution: he would call *all* the yellow-taxi companies when he needed a cab. Then he would take the first one that arrived.

Not surprisingly, the cab fleets didn't like that tactic. Though it is impossible to confirm, Garrett Camp believes his cell phone was blacklisted by the San Francisco taxi companies. "Eventually they wouldn't take my calls," he says. "I was banned from the San Francisco cab system."

Then Camp got a girlfriend.

A few months after eBay's acquisition of StumbleUpon, he sent a message over Facebook to a smart, beautiful television producer named Melody McCloskey, and—after noting that they had a vague connection because they shared the blogger Om Malik as a friend on the social network—asked her out on a date.

McCloskey, who is now the founder and CEO of the online beauty and wellness company StyleSeat, recalls being wary, but she agreed to go for coffee. Camp suggested they meet at a restaurant at eight on a Friday night. She countered that a café at six p.m. on a Tuesday might be more appropriate. He offered seven o'clock on a Thursday night as a compromise and then changed the meeting location to a bar at the last minute.

McCloskey told herself she was going for only forty-five minutes. They stayed out until 2:00 a.m. "I accidentally went out with this person on this wild date," McCloskey recalls years later. "I don't think I made it into work the next day."

Like many high-tech entrepreneurs, Camp was peculiar. McCloskey noticed that he did not particularly care about the superficialities that absorbed other people. For example, he got his hair cut only sporadically, letting it grow down to his shoulders before having it cut short. He also liked to design his own T-shirts featuring symbols such as a Necker cube, a line drawing that can be perceived in different ways. Then he would wear them out to dinner at nice restaurants. "I have no idea where he got those things," McCloskey says. "I was not thrilled by them."

He didn't like to carry cash and would come home and absentmindedly stuff an unwieldy wad of bills into his dresser, then leave it there. Though Camp was a newly minted millionaire and McCloskey at the time was scraping by as a producer for the cable news network Current TV, "I was paying for everything," she says.

The relationship posed a new set of transportation hurdles as well. McCloskey lived a few miles away from Camp, in Pacific Heights. Meeting anywhere was a hassle and Camp often wanted to get together somewhere out at night.

"The logistics of dating you are very hard," she told him once. "I can't afford to meet you all over the city. I can't keep up with your lifestyle."

To solve these mounting transportation challenges, Camp started to experiment with the city's gypsy-cab fleet—the unmarked black sedans that would approach prospective passengers on the street and flash their headlights to solicit a fare. Most San Franciscans, particularly women, would stay away from those unmarked cars, fearing for their safety or worried by the ambiguity of a cab without a running meter. But Camp found that a majority of the cars were clean and that many of the drivers were friendly. The biggest problem for these drivers was filling in the dead time between rides, when they tended to wait outside hotels. So Camp started collecting the phone numbers of town-car drivers. "At one point, I had ten to fifteen numbers in my phone of all the best black-car drivers in San Francisco," he says.

Then he started gaming the system further: texting a favorite driver hours before he needed him and telling him to meet him at a restaurant or bar at an appointed time. On another night, he rented a town car and driver for himself and a group of friends for an entire evening. It was an indulgence that cost a thousand dollars, and zooming around the city at the end of the night dropping everyone off was a pain.

And that is when the futuristic image from the James Bond movie *Casino Royale* popped into Garrett Camp's head.

Suddenly Camp was obsessed with a new notion. He frequently talked with McCloskey about the idea of an on-demand car service and vehicles that passengers could track via a map on their phones. At one point that year, Camp scrawled the word *Über*—with the umlaut over the *U*—into a Moleskine notebook that he kept to jot down new ideas and logos for companies and brands. "Isn't that pronounced Yoober?" she asked him.

"I don't care. It looks cool," he said.

McCloskey recalls that Camp "wanted it to be one word and a description of excellence" and that his musings on the word, its sound and meaning, were incessant. "What an uber coffee that was," he'd say randomly after drinking a cup. "It means great things! It means greatness!"

Camp says he contemplated calling this new service ÜberCab or BestCab and finally settled on just UberCab, losing the umlaut. (He registered the domain name UberCab.com in August 2008.) McCloskey loved Camp's endless examination of new ideas but wasn't so sure she believed in this particular one. "Sure, cabs are terrible," she said. "But you are only in the cab for eight minutes! Why does it matter?"

But Camp was certain that *he* wanted such a service. He also knew that the iPhone and its new App Store, which Apple introduced over the summer of 2008, were going to finally make the futuristic vision in *Casino Royale* practical. Not only could you chart the location of an object on a map but, since the earliest models of the phone had an accelerometer, you could also tell if the car was moving or not. That meant that an iPhone could function like a taximeter and be used to charge passengers by the minute or the mile.

He talked it over that year with many of his friends. The author and investor Tim Ferriss first brainstormed with Camp about the then-unnamed Uber at a bar in the Mission District. He thought it was a great idea, then forgot all about it. A month or two later he got a call from Camp, and when they started talking about Uber again, Ferriss was shocked. Camp "had done an incredibly deep dive into the flaws of black cars and a kind of lost utility, the downtime of black cars and taxis," he says. "It was clear that he was probably already in the top one percent of market analysts who have looked at the space."

The idea behind Uber was crystallizing in Camp's mind.

Both the passenger and the driver could have an app on their phones. The passenger could have a credit card on file and wouldn't have to travel with any pesky cash. "I bounced the idea off of everyone," Camp says. "All these ideas kept building and building."

The original idea was to buy cars, then share the fleet among his friends who were using the app. But Camp says that was only a starting point and that even back then he was considering the potential to use such a system to coordinate not just black taxis but eco-friendly Priuses and even yellow cabs.

"I always thought it could become a more efficient cab system, particularly in San Francisco," he says. He wasn't sure it would work outside the city, though. If he could get it to work in just a hundred cities, he reasoned, it could be big enough for a company that generated about one hundred million dollars a year in service fees.

By the fall Camp had more free time to work on Uber, since he and McCloskey had broken up, though they remained friends, and he was going less frequently into StumbleUpon. He recalls spending his weekends getting coffee, cruising the web, and doing research into the transportation industry and then going out with friends at night.

On November 17, 2008, he registered UberCab as an LLC in California. Soon after, hungry for some basic market research, he sent an e-mail to Ferriss asking if the author could have his assistant do some digging for him. He included a link to a wiki, an online document that they could both access. Years later, Camp reads out loud a few of the one hundred questions he laid out in the wiki.

> COMPARABLE SERVICES (five hours of research wanted). Are there any one-click luxury on-demand car services in existence? How big is the total market for on-demand chauffeured transportation?
>
> LOGISTICS AND FEASIBILITY (ten hours of research wanted). How long does it take to get a limo permit from the public utilities commission of California? What is the average pickup time in minutes when someone calls a cab—average and median—in the top ten U.S. cities? How many taxicab companies offer guaranteed pickup?
>
> CAB-INDUSTRY DYNAMICS (five hours of research wanted). What are the critical must-haves in dispatch software? How much of the dispatch process can be automated?

At the end of the e-mail, Camp wrote to Ferriss, "My goal is to be at a go or no-go decision by December 1 and to be live with five cars in January."

Camp does not recall getting much help from Ferriss's assistant, but nevertheless he plunged ahead. In December, on the way to LeWeb, a high-profile annual technology conference in Paris, he stopped in New York City. There he met Oscar Salazar, a friend and fellow graduate student from the University of Calgary.

Salazar was a skilled engineer from Colima, Mexico, the son of an agronomist (a technician who worked on farms) and a kindergarten teacher. As an aspiring entrepreneur in his early twenties, he had built a wireless mesh network over his hometown by putting Wi-Fi antennas onto electrical poles and roofs. But he never got permission, and the city shut it down. Eager for an environment more supportive of innovation, he got his master's in electrical engineering in Canada and his PhD in France, then moved to New York.

During this time, he kept in touch with Camp, and they reunited that December at a delicatessen in lower Manhattan. Camp pitched UberCab to Salazar and asked him to lead development of the prototype.

"I have this idea. In San Francisco it's hard to get a taxi. I want to buy five Mercedes," Camp said, taking out his phone and showing him a picture of a Mercedes-Benz S550, a high-end coupe that sold for around a hundred thousand dollars. "I'm going to buy the cars with some friends and we're going to share drivers and the cost of parking." He showed mock-ups of iPhone screens demonstrating how cars would move on maps and how passengers might see a town car coming toward them.

Salazar had experienced his own troubles hailing cabs in Mexico, Canada, and France and remembers telling Camp as he signed a contract, "I don't know if this is a billion-dollar company but it's definitely a billion-dollar idea." Since Salazar was in the United States on a student visa, he couldn't receive payment in cash for the job. Instead, he received equity in the fledgling startup. His stake is now worth hundreds of millions.

"It's way more than I deserved. It's more than any human deserves," he told me over breakfast at a New York City café in 2015.

UberCab was officially in development. And so Camp left for Paris and the LeWeb conference, where he was meeting McCloskey and a close friend and fellow entrepreneur—Travis Kalanick.

Every company creates its own origin myth. It's a useful tool for expressing the company's values to employees and the world and for simplifying and massaging history to give due credit to the people who made the most important contributions back when it all started.

Uber's own official story begins here in Paris, when Camp and

Kalanick famously visited the Eiffel Tower on a night after LeWeb and, looking out over the City of Light, decided to take on an entrenched taxi industry that they felt was more interested in blocking competition than in serving customers.

"We actually came up with the idea at LeWeb in 2008," Kalanick would say five years later at the same conference, citing the challenges of getting a cab in Paris. "We went back to San Francisco and we created a very simple, straightforward to us at the time, [way] to push a button and get a ride. We wanted it to be a classy ride."[3]

Like all mythologies, it is not really true. "The story gets misrepresented a lot of times." Camp sighs. "The whole LeWeb thing. I'm okay with it, as long as it's directionally correct."

Camp had previously discussed the Uber idea with Kalanick, as he had with other friends. The pair shared an enthusiasm for starting companies and solving technical problems, as well as for coining new phrases and mining the potential of words. While Camp would ruminate on the meaning and sound of *uber,* Kalanick liked to say he was "nonlucky" in his previous startup experiences. He had dubbed his San Francisco apartment, where entrepreneurs would gather to jam on new startup ideas, the Jam Pad. It was a kind of entrepreneurial safe house, a place where like-minded obsessives could gather in front of a whiteboard and debate the intricacies of building internet companies.

At the time, Kalanick was enthusiastic about Camp's notion for a smartphone-based town-car-sharing service but only mildly interested in getting involved. He had just sold a previous startup, the streaming-video company Red Swoosh, to a much larger competitor, Akamai, and was in the middle of what he later called his "burnout phase," traveling through Europe, Thailand, Argentina, and Brazil, and sizing up different career options. "Travis thought it was interesting but he was in this mode," Camp says. "He had just left Akamai and was traveling a lot and angel investing. He wasn't ready to go back in."

In Paris, they all stayed at a lavish apartment that Kalanick had found on the website VRBO. Camp was talking endlessly that week about Uber, but Kalanick had his own startup idea, which, considering everything that subsequently happened, was ironic: he was envisioning a company that would operate a global network of luxurious lodgings, identically furnished and separated into different classes, that could be leased via the internet. Frequent business travelers could subscribe to this network, rent places, and pay for them seamlessly. Riffing on the nickname of his own home, the Jam Pad, he called this business idea Pad Pass. "It was sort of a cross between a home experience and a hotel experience," Kalanick later told me. "I was trying to bring those two together." Camp recalled it too. "Travis had hacked out a whole Airbnb-like system that we were considering starting," he said. "Uber was my idea; that was his idea."

McCloskey remembers that Kalanick had reached the same conclusions as the founders of Airbnb. The internet could allow travelers to find luxurious yet cheap accommodations while also offering a far more interesting traveling experience. "He was frustrated by VRBO," she says. "Its payment service was shitty and you couldn't book it instantly like a hotel but had to e-mail back and forth. He just wanted to fix all that stuff."

Nevertheless, the conversation that week in Paris gradually came to focus more on Uber than Pad Pass. Camp was convinced that the right way to start the business was to buy those top-line Mercedes. Kalanick strongly disagreed, arguing that it was folly to own the cars and more efficient just to distribute the mobile app to drivers.

McCloskey remembers one dinner at a fancy restaurant at Paris where the debate raged over the best way to run an on-demand network of town cars. The restaurant was elegant, with expensive wine, light music, and a sophisticated French clientele. Apparently there was also paper over the tablecloth because Camp and Kalanick spent the entire meal scrawling their estimates for things like fixed costs and maximum vehicle utility rates.

"When we left that dinner, the entire tablecloth was covered in math," McCloskey says. "There was no 'Let's go to dinner and talk about life.' This was Travis's life, connecting over analytical problem-solving. That was how he connected with people." *Parisians must think Americans are the craziest people on the planet,* McCloskey remembers thinking as they left the restaurant.

On a separate night in Paris, the group went for drinks on the Champs-Élysées and then to an elegant late-night dinner that included wine and foie gras. At 2:00 a.m., somewhat intoxicated after a night of revelry, they hailed a cab on the street.

Apparently they were speaking too boisterously, because halfway through the ride home, the driver started yelling at them. McCloskey was sitting in the middle of the backseat, and, at five feet ten inches tall, she'd had to prop her high heels on the cushion between the two front seats. The driver cursed at them in French and threatened to kick them out of the car if they didn't quiet down and if McCloskey didn't move her feet. She spoke French and translated; Kalanick reacted furiously and suggested they get out of the car.

The experience seemed to harden their resolve. "It definitely lit a fire," McCloskey says. "When you are put in a situation where you feel like there's an injustice, that pisses Travis off more than anything. He couldn't get over it. People shouldn't have to sit in urine-filled cabs after a wonderful night and be yelled at."

That cantankerous Paris taxicab driver may have left an indelible mark on transportation history.

By the time they got back to San Francisco, Kalanick was ready to get more involved, at least as an adviser, and Camp was ready to listen to him. A few weeks into 2009, after their trip to Washington, DC, to see Barack Obama's first inauguration as president, Camp called Kalanick. He was about to lease parking spaces in a garage near his home on Hawthorne Street in San Francisco for the fleet of Mercedes he was still determined to buy.

Kalanick counseled him against it one last time: "Dude, dude! You don't want to do that!"

Camp finally gave in and ended the ongoing debate; he never signed the lease and never purchased the cars. Instead of buying a dozen flashy Mercedes, Camp, along with Kalanick, would pitch the app to owners and drivers of town cars.

Kalanick would brag a few years later, in one of our first interviews: "Garrett brought the classy and I brought the efficiency. We don't own cars and we don't hire drivers. We work with companies and individuals who do that. It's very straightforward. I want to push a button and get a ride. That's what it's about."

Despite the initial burst of creativity, Uber gestated slowly in 2009. The company's founders still viewed it as a side project and were consumed by other things. That April, eBay spun out StumbleUpon, amid declining traffic and debate about its future, and the newly independent company received an injection of fresh capital from Camp and a group of investors. Camp resumed his position as CEO.[4] Kalanick, meanwhile, continued to travel, invest in startups, and serve as an adviser to other San Francisco entrepreneurs.

For three Mexican developers, though, Uber was a full-time vocation. In New York, Oscar Salazar took Camp's ideas and started to design the mechanics of the service. Looking for help, he subcontracted the duties of programming the first Uber dispatch system—the algorithm that would match passengers with the closest vehicle—to a hardworking friend from Colima named Jose Uribe and his girlfriend at the time (now his wife), Zulma Rodriguez.

The couple, both engineers, tended to immerse themselves totally in projects, working from morning to night in Uribe's childhood bedroom at his parents' home in Colima. Salazar had asked them to help with a variety of his projects, including a text-based tool for alerting patients that it was time to take their medications. Now he had a new job for them. At first Uribe asked for payment in cash; Salazar convinced him to take equity as well. That small stake is now worth millions. "I try not to think about that," Uribe said in an interview. "I don't want it to affect me."

Uribe and Rodriguez worked on Uber almost exclusively from February to June in 2009. They sketched out the dispatch algorithm on paper and discussed it over the phone with Salazar in New York before starting to code in the open-source computer languages PHP, JavaScript, and jQuery. Ideas that are still part of the Uber service were codified back then; the fare for a ride was determined by adding a rate per kilometer with a rate per minute. The biggest challenge, Uribe recalled, was "locating the closest vehicle and optimizing it so the process would be fast."

In that first version of Uber, passengers could request the vehicle by sending their address via a text message to a special telephone number known as a short code. The dispatch software would then locate and relay the message to a nearby driver. This first SMS-based dispatch did not work well, in part because if passengers made a mistake inputting their address, the driver wouldn't be able to find them. The engineers also created the option to request a vehicle through the UberCab website, an idea the company quickly abandoned since few people were surfing the web while they were on the street looking for a cab.

The group also worked on a version for the iPhone. Camp had sent Salazar a copy of a February 2009 issue of *Wired* magazine with a cover story entitled "Inside the GPS Revolution." The article included short profiles of location-aware apps "that deliver the hidden information that lets users make connections and interact with the world in ways they never imagined."[5] Camp suggested Salazar call one of the companies profiled in the magazine for help.

Salazar ended up selecting the maker of an app called iNap, which let train travelers specify the geographic location where they wanted their iPhone alarms to wake them up. Salazar wrote the creator of that service, a Dutch UI designer named Jelle Prins, through his website and hired him and his partner Joris Kluivers to develop Uber's first app for the iPhone.

By fall there was a working prototype. In September, Camp and Kalanick attended the Lobby, an annual off-the-record networking event in Hawaii hosted by the venture capitalist David Hornik, and started quietly pitching the concept to entrepreneurs and investors. Kalanick was getting excited and putting in a few hours a week on it. Around that time, Camp introduced Salazar to Kalanick over e-mail. "Garrett introduced Travis as an adviser to the company," Salazar says. "He didn't want to get fully involved, but Garrett was trying to convince him. Garrett knew he could be perfect for it."

A few weeks later, Camp and Kalanick met Salazar in New York's East Village and tried the app for the first time in authentic conditions. They hired a few random black-car drivers, who likely did not suspect that history was being made, gave them iPhones with the app, fanned out across lower Manhattan, and tried to summon the cars from various locations via smartphone.

It was buggy and barely worked. Afterward one driver told Camp as he returned the iPhone, "Well, that was really hard."

The group went for pizza on Prince Street in SoHo and talked about what needed fixing. They were deflated about the test but excited about the concept. It was real now; it was tangible. When it worked, they could see the car moving across a map toward them, like Camp had originally envisioned, like James Bond in *Casino Royale*. Back in California a few weeks later, Camp and Kalanick met the founders of a Palo Alto mobile application consulting firm called Mob.ly and reassigned the development of the iPhone application to them.

By early 2010, Kalanick and Camp agreed on one thing: they both wanted to use Uber but neither of them wanted to run it. Camp was an inventor and loved being around the creation of an idea. He also had his hands full with StumbleUpon. Kalanick still valued his freedom and the opportunity to advise many startups at once. If he was going to devote himself to a new idea, it had to be a big one. This was only a limo company, a new way to shepherd relatively affluent users around the city in style.

So on January 5, 2010, Travis Kalanick Tweeted in the peculiar shorthand common to the 140-character messaging service:

> Looking 4 entrepreneurial product mgr/biz-dev killer 4 a location based service...pre-launch, BIG equity, big peeps involved—ANY TIPS??

Halfway across the country, in Chicago, Illinois, a twenty-seven-year-old General Electric employee named Ryan Graves sent the single most lucrative Tweet in internet history:

> @KonaTbone heres a tip. email me :] graves.ryan[at]gmail.com.

Graves was not really the Silicon Valley type. He was tall with a sunny disposition and nearly perfect hair, resembling "the star of a cigarette ad from the 1950s," as one investor put it.

His childhood in San Diego was quintessentially American; his father had been a radio-advertising salesman while his mother tended to the family and ran a women's Bible study network. He graduated from Miami University in Ohio with a bachelor of arts in economics in 2006 and was not initially interested in the field of technology. But he seemed to have an endless appetite for turning himself into an expert on any topic that fascinated him. So far, that had included European soccer, fly-fishing, motorcycles, and great surfing spots. Now it was the lucrative and energizing internet economy, and he wanted to find a job in it.

While he went through GE's management-training program, Graves completed a business-development internship at the location app Foursquare. He had tried to develop his own social app, with little success. Though he was technically in General Electric's leadership training program, he wasn't there a lot. "You can come in at ten a.m. and leave at four p.m. and no one knows," he says. "I was putting in very little time at GE while getting a very high ranking."

Kalanick was interested enough to meet with him, so Graves snuck out of a GE training class in Crotonville, New York, and drove an hour to New York City to meet Kalanick at a SoHo coffee shop. They talked for more than two hours and Kalanick showed Graves the prototype iPhone app.

Graves was intrigued. This was the opportunity to run something by himself. It was also a position working with some very connected Silicon Valley entrepreneurs and most likely would allow him to stick a foot in a much bigger door. Plus: "I don't think there was anyone else competing for the job," he says.

Two weeks later, Graves moved to San Francisco, while his wife, Molly, a teacher, stayed in Chicago until the end of the school year. He had prepared a slide deck with his vision for the car-sharing service. Kalanick edited it, and together they presented it to Camp.

Graves tried the app again, this time in San Francisco, where a few drivers were already testing out a beta version of the service. "It didn't work for shit," he remembers. The wireless coverage, provided for the iPhone back then exclusively by AT&T, was horrible, and the Uber app's use of GPS quickly drained the iPhone's battery. "This doesn't really work," Graves recalls telling Kalanick and Camp, "because I thought you told me that it does."

Uber didn't yet have an office, so Graves worked from a hotel and from cafés around the city and started to get to know his fellow entrepreneurs. One of his first meetings: Brian Chesky, who met him at a café, Rocco's, in the South of Market neighborhood. Graves wanted advice on his compensation negotiations with Camp and Kalanick. "I remember him pitching it a little like Airbnb for cars," says the Airbnb CEO. "It sounded really cool, but how big was the market for black cars?"

Graves was introduced to his first engineer at a bar. Camp had invited another classmate from graduate school, Conrad Whelan, to join UberCab after Whelan told him he was finally ready to leave Calgary.

Now that there were two employees, they needed an office. Graves had met the founder of the online-travel startup Zozi over Twitter, and Zozi happened to have a small unused windowed conference room in its offices across the street from the iconic Transamerica Pyramid. So the UberCab staff set up shop there, on the second floor, sitting in chairs at a square desk that was wedged against the wall.

The company hoped to launch the service to the public that summer.

Whelan worked with Salazar in New York, with Uribe and his wife in Colima, and with the Mob.ly team in Palo Alto to add features to the app, like a way for users and drivers to register for the service. Meanwhile, Graves, the CEO, and Kalanick, an adviser now spending about twenty hours a week on UberCab, cold-called and visited San Francisco town-car fleets and pitched the service to the owners. "It was old school dialing for dollars," Kalanick would later say.[6] "A third of the calls, you know, basically I got hung up on before I got to the core pitch. A third of the calls they heard for about a minute and a half and then I got hung up on. And a third were like, this is interesting."

In May, Mob.ly was acquired by Groupon and announced it would shut down its ongoing projects. It was almost disastrous for the fledgling UberCab. Graves had to beg Mob.ly to finish stable versions of the apps for riders and drivers. The company agreed, and in the first week of June 2010, UberCab's apps went live in Apple's iOS App Store. An idea that had popped into Garrett Camp's head a year and a half before now quietly launched in the city of San Francisco, right as the smartphone revolution began to gather momentum.

Now the company needed real capital.

What happened next would define the careers of hundreds of

Silicon Valley financiers. None of them knew they were about to make the most important decision of their professional lives.

Most of Silicon Valley's best and brightest passed on the deal, just as they had with Airbnb. They said no because Ryan Graves wasn't experienced enough or because the two founders weren't involved enough or because they saw the concept as an extravagant indulgence for wealthy urbanites. Some said no because they had worked with the combative Travis Kalanick before at his previous companies and didn't want to deal with the aggravation again; others because they knew the company was going to run headlong into a hostile tangle of city and state transportation laws.

They said no and then later, after the company hit it big, insisted that the original e-mail had gone to their spam filters, or that they had simply overlooked it, or that they had been on vacation at the time. If they were being honest, they spoke about the missed opportunity in hushed tones and with pained expressions.

Another reason they said no was that Uber looked nothing like what it would later become. This is the cruel reality of startup investing—financiers are betting on a future they can't see. At the time, Uber had been in the iOS App Store for all of two weeks. Ryan Graves and Travis Kalanick had succeeded in recruiting about ten town-car drivers in San Francisco. The service was facilitating ten rides a weekend, with most of those probably taken by Uber's own employees, founders, and friends. The company wielded only one statistic in its original pitch: half of the people who downloaded the app and signed up actually tried it out and went for a ride.

Both Camp and Kalanick were well connected, so they were able to avoid the awkward groveling that the Airbnb founders had endured the previous year. They kicked off the process simply by contacting a friend, Naval Ravikant, who had created an e-mail network of SEC-accredited investors called AngelList. Kalanick had been informally talking to Ravikant about becoming a partner at AngelList, and Ravikant offered to help Kalanick try out the service to reach out to some top investors.

A message was sent out to the one hundred and sixty-five investors on AngelList on June 17, 2010. "UberCab is everyone's private driver," the e-mail said. "We're solving the taxi scarcity problem with on demand private cars via iPhone and SMS." The e-mail said that Camp was the founder and investor, that Kalanick was a "mega-adviser" and would be an investor in the seed round, and that Tim Ferriss was an adviser and investor. It introduced Ryan Graves as CEO, who testified in the e-mail that he had hustled his way into a business-development internship at Foursquare.

"Should we make an intro?" the e-mail asked investors.

Ravikant says that one hundred and fifty of the one hundred and sixty-five investors didn't respond. One investor actually unsubscribed from the list after getting the e-mail.

Even those renowned for backing nearly everything (members of the "spray-and-pray" school of seed investing) declined. Ron Conway, the "godfather" of Silicon Valley, famous for backing the holy trinity—Google, Facebook, and Twitter—passed on the deal. "This one looks like it will be a fight in every city," he sagely e-mailed a fellow investor. Dave McClure, who would go on to create the startup accelerator 500 Startups, said that he really didn't know Ryan Graves well enough to invest.

Investor Bill Gurley, a partner at Benchmark who would hitch his ride to the Uber rocket a year later and was watching the taxi market closely, took Kalanick and Graves out to dinner in early July to Absinthe, a restaurant in the Hayes Valley neighborhood of San Francisco. But Benchmark didn't usually invest in seed deals and he couldn't get his partners to make a commitment so early.

Then there were those who said yes.

The Philadelphia-based venture capital firm First Round Capital led the round with a $600,000 investment. Rob Hayes, a partner at the firm, had backed StumbleUpon and saw Camp Tweeting about UberCab. "I'll bite—what's Ubercab?" he e-mailed him.

Camp sent Graves to pitch Uber at First Round's San Francisco offices, and the partners voted unanimously to approve the deal.

Rob Hayes spent so much time closing it over the Fourth of July weekend that he angered his family. "I was betting on Ryan Graves and Garrett Camp," Hayes says. "I didn't meet Travis until the first board meeting."

There were a dozen other seed investors. Chris Sacca, a former Google executive with a taste for embroidered cowboy shirts who had just bet big on Twitter, heard about it over a sushi dinner in San Francisco with Kalanick, Camp, McCloskey, and Oscar Salazar.

"There's this guy, an investor, he's crazy," Salazar recalls being told beforehand. "We're going to have dinner with him. Don't mention anything about Uber. Just say the name; he doesn't have to know what we're working on."

When they finally dropped a hint, Sacca immediately bit. He knew both Camp and Kalanick well enough to sense that together they could accomplish something special, and he wrote a three-hundred-thousand-dollar check almost on the spot. "This is the one I got really fucking right," Sacca says.

Others were similarly impulsive. Mitch Kapor, creator of the early 1990s productivity tool Lotus Notes, was furious at himself for asking for his money back from the failed podcasting company Odeo right before it morphed into Twitter. So he was aggressively pursuing all leads. "I'm in," he said to Camp, whom he had backed in StumbleUpon. "If you don't let me into the deal I'll kill you."

Jason Calacanis, a blogger and founder of internet media start-ups, was friendly with Kalanick and invited him to pitch to a group of investors at his own event in San Francisco, the Open Angel Forum. Kalanick found a few willing backers there, including Calacanis himself, who for the next ten years would revisit his investment decision on various podcasts, in blogs, and on the Q-and-A website Quora.

But as much as they might want to cast their decisions as intuition, most of the early investors would have to bow to that unpredictable Silicon Valley divinity—luck. "It was controversial to invest in the company," says Alfred Lin, then the chief operating officer at the online shoe retailer Zappos. When Lin heard about the deal, he doubted Uber could ever work back at his home in Las Vegas. He had serious misgivings. "I just thought that founders who are passionate about the idea would run the company," he says. But he ponied up anyway after trying out the service in San Francisco and deciding he'd rather not be left out.

David Cohen, co-founder of the Colorado-based startup school Techstars, got a chance to invest only because of a geographic accident. Ryan Graves had to fly to Chicago that summer so he and Molly could ferry their possessions back across the country. On the drive to San Francisco, he made call after call, pitching UberCab so frequently that Molly could recite the spiel word for word. They happened to pass through Boulder, and Graves stopped to meet Cohen, who liked the pitch and ponied up $50,000. "Luck is a part of this game," Cohen would later write in a blog post about his decision.[7]

Even though they got lucky, some of the earliest Uber investors had to live with the fact that they could have been even luckier. Ravikant of AngelList planned to invest $100,000 but waited until the end of the fund-raising process to avoid the appearance that he was favoring certain AngelList deals over others.

When Ravikant finally made an offer, Graves said that he no longer had room in the round. Ravikant begged and eventually got to put in $25,000. It's still by far the best investment he's ever made (current value: over $100 million). "I don't fixate on it," Ravikant says. "I've made peace with the fact that Silicon Valley is so random. You have to make peace with it or otherwise you'll never get a good night's sleep in this town."

Now Uber had $1.3 million in the bank, a $5.3 million valuation, an office (small and crowded), and a product (buggy). It was finally looking like a real startup. Uber's founders and investors told their influential and affluent San Francisco friends, and word began to spread. On July 5, the blog *TechCrunch* wrote its first story about the app: "UberCab Takes the Hassle Out of Booking a Car Service."

"Of course, convenience has a price," noted the article's author, Leena Rao. "You may pay anywhere from one and a half to two times the price of a cab fare (but two times less than a traditional car service fee). But you are receiving better service, a nice black limo and an on-demand solution."[8]

With the wind now blowing softly at his back, Graves started to build his staff. Ryan McKillen, one of the new hires, attended Miami University in Oxford, Ohio, a year ahead of Graves and they had several friends in common. They had hung out earlier that year while Graves was in San Francisco without his wife. Then, fortuitously, as it turned out, the accounting startup where McKillen rather enjoyed working imploded. So Graves hired him. (Because the two men have the same first name, colleagues took to calling them by their initials, a practice at Uber that endures today.)

On his first day in the office, McKillen noted the table had piles of programming books in pristine condition and a well-worn Spanish/English dictionary. (The engineers had been trying to translate some of the instructions around the code written by Jose Uribe.) McKillen asked Conrad Whelan why the dictionary was there and later he would enjoy recalling Whelan's response: "Well, Ryan, because the code is written in Spanish. Welcome to Uber."

Austin Geidt's route into Uber was even more unlikely. She grew up in Marin, north of San Francisco, and attended the University of California at Berkeley, and while in college, she suffered from an addiction to heroin. When she finally recovered and graduated, she was adrift, insecure, and desperate for a job. After she applied for a barista position at the Peet's Coffee in Mill Valley and was turned down, she spotted a Tweet by Jason Calacanis about Uber. She followed a few links, then sent an unsolicited e-mail to Ryan Graves. He hired her as a marketing intern.

By her own account, she struggled to fit in. On her first afternoon the whole company went to Kalanick's apartment for a multi-hour jam session about the future of the company and the meaning of its brand. During the discussion, which lasted all evening, Geidt noticed that Kalanick, who paced the room, seemed to be really in charge. She found the experience incredibly stressful. "It was a bit overwhelming for me," she says. "I had terrible impostor syndrome and I hadn't even done anything yet."

Over the next few months she was certain she was about to get fired. She had a poorly defined role that at one point required her to hand out flyers for UberCab in downtown San Francisco. Even the trivial act of writing an e-mail would send her scurrying to her older siblings for help and advice. At one point, Graves recalls, he counseled her in the stairwell to the office, with Geidt in tears. Instead of firing her, though, Graves gave her time to find her footing. Later he dismissed his first driver-operations manager and installed Geidt in the role. She would become one of the most important execs in Uber's early history.

By the fall of 2010, San Francisco was starting to notice Uber. The service was exceedingly viral; a user stepped out of a town car and walked into a bar, and suddenly his or her friends wanted to know everything about it.

Limo and town-car drivers were also intrigued. They started showing up in the Uber office, one by one. Conrad Whelan recalls watching Graves give a driver the pitch and train him to use the app. Afterward, the driver laughed. "Oh, you guys are going to make a lot of money." That's when Whelan canceled his vague plans to return to scientific research.

Uber was turning into something special, generating positive word of mouth and even minting a few local celebrities. Sofiane Ouali was an immigrant from Algeria who had arrived that fall in San Francisco. He knew five languages and had a background as a petroleum engineer, but he found the easiest way to get on his feet in a new country was to start driving. The owner of a fleet of black town cars who was curious about the new app charged Ouali with trying the service, and he gave him the runt of his fleet—a white 2003 Lincoln Town Car. Pretty soon early Uber riders were Tweeting about the magical appearance of a car they dubbed "the unicorn."

"I saw the way people talked about Uber and all the positive words they used about it, and I knew it was going to be big," says Ouali, who became closely acquainted with early Uber riders like Kalanick, Camp, Geidt, and Brian Chesky.

Others were noticing the hubbub too. That fall, complaints by taxi drivers and yellow-cab fleet owners about a new unlicensed competitor started to pour into the offices of city and state regulators. They claimed it was illegal and should be shut down. So on October 20, 2010, four months after the launch, when Graves was at a board meeting at First Round Capital with Travis Kalanick and Garrett Camp, four government enforcement officers walked into the tiny UberCab office.

Two were from the California Public Utilities Commission, which regulated limousines and town cars, and two were from the San Francisco Municipal Transportation Agency, which regulated taxis.

The plainclothes officers flashed badges, and then one of them held up a clipboard with a cease-and-desist letter and a large, glossy head shot of a smiling Ryan Graves. Waving the photograph around the room, he demanded: "Do you know this man?"

CHAPTER 3

THE NONSTARTERS

SeamlessWeb, Taxi Magic, Cabulous, Couchsurfing, Zimride

Everyone failed in the taxi industry. The fleet owners failed. The drivers failed. Riders spoke clearly. Some people chose to listen and some didn't. I was part of it, and I accept it.

—Thomas DePasquale, founder of Taxi Magic

Years before UberCab started ferrying people around cities and AirBed & Breakfast began offering spare couches and bedrooms, a young lawyer named Jason Finger sat down one night at his office in New York City and decided to solve the pesky problem of what to order for dinner.

It was 1999, the height of the frenzy of the first dot-com boom. Finger was fresh out of law school and working at the law firm O'Sullivan, Graev, and Karabell. For some reason he had taken it upon himself to walk around the floor every evening and collect dinner orders from his other late-working junior colleagues. Calling in these orders, coordinating payments, and then organizing the incoming food-delivery people, who all appeared in the lobby wielding soggy plastic bags at the same time, was about the nightmare you might expect.

So Finger and a friend decided to come up with a solution. They built a food-ordering website that catered to law firms and investment banks. They called it SeamlessWeb.

SeamlessWeb launched in April of 2000 and ran right into the teeth of the dot-com bust. Finger raised less than half a million dollars, paltry by the overcaffeinated standards that came later, but the service caught on quickly with the employees at several high-powered law firms and investment banks. SeamlessWeb contracted with hundreds of Manhattan restaurants and gave its corporate customers and their employees a way to browse menus and place orders over a website, expense meals to the company, and coordinate the flurry of deliveries.

The business, headquartered in midtown Manhattan on the corner of Thirty-Eighth Street and Sixth Avenue, grew briskly. Restaurants appreciated the increased volume and the companies loved how it tamed the chaos of monthly expense reports.

If there's a forefather of the crowded family of on-demand delivery startups that now jam the tech hubs of the United States, Asia, and Europe, it's SeamlessWeb. Finger was among the first to see that the internet could do more than connect people with information and one another in a purely digital realm; it could also efficiently move physical things in the real world. He understood that if it worked for food, it could work for other things. And he drew up plans to take advantage.

Seamless Meals, as he called it, was going to be one service. He also had another idea, which he dubbed Seamless *Wheels*.

The notion was to create an easy way to book and expense town cars, the same way the company had made it easy to order food. Finger registered the URL SeamlessWheels.com in 2003 and over the next few years started introducing the service to blue-chip law firms like Dewey and LeBoeuf, White and Case, and Debevoise and Plimpton.

The investors he approached about Wheels were wary.

"Every institutional investor I spoke with was like, 'Black cars are niche, it's only New York City, it's only bankers, there are long-standing relationships with companies, there is no opportunity in the consumer market,'" Finger says.

At one point the transportation coordinator at a law firm suggested that Finger tread carefully, because the rumor was that the Russian Mafia was involved in New York City's black-car business. The Italian Mafia will kill you, the saying goes; the Russian Mafia will let you live and kill your entire family. Finger dismissed the warning.

Then one day, he arrived at his office and picked up a voice mail. The message was from a man who did not identify himself and left no return number. It has long since been deleted, but both he and his wife, Stefanie, who also worked at SeamlessWeb and listened to the message, recall it the same way:

> *Jason, we understand you've been pitching a car service to large enterprises in the New York City area. We don't think that would be a good idea. You've got such a beautiful family. Why don't you spend more time with your beautiful baby daughter? You've got such a good thing going with your food business. Why would you want to broaden into other areas?*

The message was "a slap in the face," Finger says. He suspected it was from one of the long-standing town-car companies that had profitably served the banks and law firms for years. They weren't eager to see an online intermediary come between them and their clients. Stefanie recalls being frightened by the message and said "just the thought of someone following us home from work was super-intimidating."

For the first time Finger wondered whether the car business was worth it. Even aside from the veiled threat, Seamless Wheels could hurt Seamless Meals. If a car kept a senior bank executive waiting at an airport, it would jeopardize the Seamless brand. This was before the era of the smartphone, and there were few ways to coordinate drivers on the road to ensure that clients got a smooth pickup experience. Then there was the reality that investors simply weren't that excited about the car-service concept.

Seamless Wheels continued working with the same law firms for a few more years, but after that voice mail, Finger largely gave up on developing it further. The food business grew and expanded to serve regular people at their homes as well as companies. In 2006, the food services company Aramark acquired SeamlessWeb and put pressure on Finger to focus on the food business, which was growing rapidly outside of New York City. Eventually he shut down Seamless Wheels.

The story has a happy ending, though. Finger raised private equity and spun SeamlessWeb out of Aramark in 2011, then shortened its name to Seamless. Two years later, it merged with a newer and smaller rival, Grubhub, and is now the leading online food-delivery company in the United States.

But Finger still thinks about the town-car market and has watched Uber's early growth with admiration and even a little jealousy. He now believes Seamless Wheels was too early and couldn't have succeeded before the era of smartphones and ubiquitous text messaging. "I look back at my life and certainly I have regrets," he says. "The car offering is not one of them. Maybe I am rationalizing because it has been a huge opportunity. But a lot of things from the timing standpoint just didn't seem to fit."

Though Seamless Wheels was short-lived, it demonstrated an incontrovertible fact: ordering cabs and submitting those tiny paper receipts for reimbursement was an expensive, time-consuming scourge in the business world and an obvious problem that technology could solve. Others noticed this as well, and in 2007, a wealthy

Virginia businessman named Tom DePasquale decided to do something about it.

His company was called Taxi Magic. Just as the search engine Alta Vista preceded Google, and Myspace dominated before Facebook, Taxi Magic would become the highest-profile precursor to Uber; the company was the first to seize, and squander, the opportunity to revolutionize the taxi industry.

In the late 1990s, DePasquale had founded a company called Outtask that made an online tool, Cliqbook, that allowed workers to book and manage their online air travel. In 2006, Concur, one of the most popular makers of expense-account software, acquired Outtask. DePasquale became an executive vice president and major shareholder of Concur and was thus in a perfect position to see a similar opportunity in for-hire vehicles, which accounted for 10 percent of all corporate travel. The following year he started a company called RideCharge with a longtime collaborator named Sanders Partee and a young Russian engineer named George Arison.

RideCharge's original apps for the BlackBerry, Windows Mobile, and the Palm smartphone allowed riders to enter the amount on the taximeter into their phones and pay the fare automatically by credit card. It allowed drivers to sidestep the dreaded knuckle-buster—those manual credit card readers that *kerthunk* back and forth over a carbon copy of a receipt that were used in cabs for far too long. RideCharge's offices were in Alexandria, at the base of the Woodrow Wilson Bridge.

When Apple's App Store opened with the introduction of the iPhone 3G in June 2008, the startup introduced an iPhone app called Taxi Magic, and soon after, it became the new name of the company. The app flourished, with tens of thousands of downloads a day. Users could select a yellow-taxi company in their city and actually summon a cab and pay the bill on their phones.

But Taxi Magic didn't really disrupt the taxi industry so much as try to work within the straitjacket of the industry's existing technology. Taxi Magic integrated with the software of the major taxi dispatch companies, like Mobile Knowledge and DDS Wireless, which were widely used by yellow-cab fleets at that point. As a result, the company could not display the icons of cabs on a real-time map, as Uber later would, because the location data in the dispatch systems wasn't precise enough. Instead, the app had a text-based status-alert page that updated with information like the name of the driver and an estimate of how far away his car was from the waiting passenger.

Taxi Magic expanded quickly to twenty-five cities during 2008, two years before the launch of UberCab in San Francisco. Concur was a major investor and promoted the service to its corporate customers. The app is an "on-demand cab service from your iPhone at the push of a button," *TechCrunch* wrote in a positive review that December.[1]

Instead of signing up drivers, George Arison and his team visited the major cities and sold the service to the fleet owners. So Arison got to know the taxi industry well. "It was an insane process," he says. "In Seattle the cab company didn't even know what a modem was, whether they had one, and, if so, where it was."

There were other problems in the world of taxis aside from general technical cluelessness. The drivers were often at war with their fleet owners over wages and employment status. The fleets were at war with one another for market share in each city. No one cared much about riders because the companies didn't have a permanent relationship with them (when passengers are standing on the side of the road hailing a cab, all taxi companies are equal). There were no penalties for bad service; as long as a driver paid the company somewhere between one hundred and two hundred dollars to take a car out for twelve hours, the fleet owner was happy—even if the cabbie did nothing but speak to friends on the phone while driving like a maniac.

The system was hopelessly broken. When a rider requested a car with the Taxi Magic app, the existing dispatch system would assign that ride not to the closest available car but to the driver in the area who had been waiting for a passenger the longest. And there was no fidelity; in order to capture the more lucrative fare, a driver on his way to pick up a Taxi Magic user might instead eagerly veer across three lanes of traffic to pick up a businessman with a suitcase who might be going to the airport. The Taxi Magic customer would be waiting forever.

The taxi fleets were reluctant to change any of this. "No technology could solve for the fact that there was resistance among taxi companies and drivers for this very basic change to the way they ran their business," Tom DePasquale says.

He is not particularly proud of what happened next. In the summer of 2009, Taxi Magic was the object of a prolonged courtship by the Silicon Valley investor Bill Gurley, a partner at the premier venture capital firm Benchmark Capital. An original backer of the online reservation company OpenTable, Gurley, who stands six feet nine inches tall, had been looking for a similar car service that could impose simplicity and efficiency on the archaic world of ground transportation.

George Arison recalls Gurley sitting in their Virginia office many times over the course of several weeks, poring over spreadsheets, talking to Partee about the taxi industry, and negotiating with DePasquale about investment terms.

Eventually Gurley made a verbal offer: an $8 million investment in Taxi Magic at a $32 million valuation. It was a chance to enlist one of the internet's most forward-thinking investors. But DePasquale, who was chairman of the company and basically its CEO, although he didn't formally hold the title at the time, declined it.

In part there was a philosophical difference. Gurley believed that Taxi Magic had potential but didn't think it had gotten the product quite right. He argued about the need to move swiftly from yellow cabs, where local governments set fares, and into the black-car and limo markets, which were less heavily regulated. They even talked about names for the new service, such as Limo Magic.

But DePasquale believed that change would have to come from

inside the taxi industry. He had also enjoyed plenty of success starting and selling his previous companies and wasn't so sure he wanted to take the advice from a West Coast venture capitalist whose investment would give him major influence over the Taxi Magic board of directors. Back then, DePasquale took pride in rejecting Gurley's overtures. "You will be the company that said no to Benchmark!" DePasquale told his senior executives, according to Arison.

Years later, the shrewd Gurley said that he walked away from Taxi Magic with a favorable impression of DePasquale. He says that although he offered to invest, he wasn't thrilled with how heavily the company depended on taxi fleets and their balky dispatch software or the fact that Concur held a 20 to 30 percent stake in Taxi Magic. "If Tom had been willing to be CEO, I would have tried harder," he adds. (DePasquale would take on the role a few years later.)

Considering the later breakaway success of Uber and its most influential investor—Bill Gurley—history doesn't reflect well on DePasquale's choices, and he knows it. "We probably could have and should have done business with him," he says now.

But he also defends the decisions he made to work within the confines of the taxi industry. "We bet the regulatory environment would hold," he says. "There was rationale behind the bet. Medallions were worth a couple million in some cities. Almost every city had police that enforced the taxi laws."

It was a failure of imagination; he couldn't envision a startup hacking century-old laws and getting away with it. "The rules have changed," DePasquale says. "The rules of what you can say to the press, the rules of fund-raising. The rulebook Uber followed is very different than the rules I was taught."

DePasquale is in his midfifties now and a wealthy man from his many successes. I had tried unsuccessfully for months to reach him for his recollections of this significant misstep. He finally called me back after Taxi Magic, which was swamped by the Uber onslaught and changed its name to Curb in 2014, was sold for a pittance to Verifone, a provider of payment terminals in cabs.

"There were a lot of other operational mistakes but they are minor compared to being on the wrong side of a bet, in an industry that was still in denial," he told me. "The industry had no appetite for even an iota of change. Ultimately as chairman and founder I should have understood that better. I can certainly waste your time and mine with what I looked at each quarter, but by that time it was too late. We were better off starting over than trying to turn the ship around.

"You can't spank me any harder than I spank myself," he continued. "Everyone failed in the taxi industry. The fleet owners failed. The drivers failed. Riders spoke clearly. Some people chose to listen and some didn't. I was part of it, and I accept it.

"There is no bitterness," he said, speaking in an uninterruptible stream. "I made lots of money in the industry. As long as you portray it like that. Write whatever you want. I'm not bitter. Uber made a risky bet that has paid off unbelievably. If you are going to lose, you might as well lose to the most successful company ever, with Bill Gurley involved. They are pretty good people to lose to."

Of all the random companies that tried to beat Uber to the transportation revolution, the unlikeliest emerged from the U.S. electronics chain Best Buy. In 2008, after Seamless Wheels had died and just as Taxi Magic was pulling out of the garage, the electronics retailer opened an in-house incubator for new business ideas. Store employees around the country were encouraged to step forward and share their startup dreams. If an employee's idea was selected, he or she got to leave the showroom floor and live and work for two months in the Park La Brea Apartments in Los Angeles. Best Buy optimistically called the project UpStart.

The UpStart program, like many such fashionable corporate initiatives, lasted only a year. No company-saving projects resulted, but an interesting one did emerge. That spring, a Los Angeles–based technician in the Geek Squad, the Best Buy unit that visits people's homes and assembles their electronics, suggested there should be a way for customers to watch an online map and see the position of a Geek Squad van that was headed to their home. The technician, Daniel Garcia, was invited to join the program to work on the idea and was assigned two interns.

The program was nine weeks long, and halfway through, Garcia and his colleagues realized their idea wasn't all that interesting. The creator of UpStart, an IBM veteran named John Wolpert, suggested they apply the technology to the taxi industry and allow cabs to be tracked on a map in the same way. Later, one of the interns, a USC graduate named Tal Flanchraych, was talking about the new Scrabble-like app Scrabulous and came up with a riff on the name: Cabulous.

They worked on Cabulous for the next few weeks, and Wolpert recognized the potential. He asked his bosses to spin the project out of the retailer. Best Buy's executives were consumed with the mounting economic crisis and happily let it go, declining even to take equity.

Wolpert set up Cabulous in San Francisco at a startup incubator called Pivotal Labs, hired a developer, and started working on an app for smartphones. He also started cultivating drivers. The city's largest taxi fleet, Yellow Cab, had a ten-year technology contract with one of the old-school dispatch companies, and another fleet, Luxor, was using Taxi Magic. But two other taxi companies, DeSoto and SF Green Cab, allowed the startup to pitch its drivers directly. Wolpert remembers long hours sitting in the front seats of cabs, learning the trade and coming to love the city's grizzled taxi drivers, who had waited years for their medallions and were clinging to a reliable stream of income. "There were so many cool guys," Wolpert says. "Old Harry Chapins, still driving. Just salt of the earth."

Wolpert was conceiving of a service that empowered those yellow-cab drivers. It would make the traditional taxi businesses more efficient and help drivers boost their earnings. This was his fatal mistake. If Seamless Wheels suffered from bad timing and Taxi Magic from stubbornness, Cabulous was doomed by civility.

"I tried to be the nice guy," Wolpert says, sitting in my office in San Francisco in early 2016, staring out the window on a rainy day and watching the Ubers pass by. "I was very into the win-win in those days. To a fault. I've learned a lot about negotiation since then."

The app debuted on the App Store in the fall of 2009, more than six months before UberCab, and offered some of the elements that would later make Uber special.

Unlike Taxi Magic, Cabulous showed the images of cabs on a map, and riders could either hail them electronically or call the fleet's dispatch number. (They could also look for their favorite driver and summon him specifically.) There were a few bells and whistles too. When users pulled up the app, they heard the sound of a car door opening and shutting and a jet engine taking off. It was totally frivolous, but years later, Wolpert plays the old sound and smiles.

Unlike Uber, Cabulous did not automatically facilitate payments; riders still had to manually pay the driver based on the fare listed on the meter. And at first, Cabulous didn't give iPhones to drivers. Wolpert had informally polled the city's cabbies one afternoon by giving away doughnuts and coffee at Bob's Donuts on Polk Street and concluded that a good percentage already owned iPhones. He didn't realize that many of those phones were jail-broken; that is, altered so they could work off AT&T's then-shaky wireless network. As a result, the app performed poorly or not at all on those phones.

The biggest problem was that Cabulous didn't control its supply of drivers or its fares and couldn't grow its fleet to keep up with demand. So when cabbies got busy on weekend nights and had plenty of street hails, "they just wouldn't turn on the app," says Tal Flanchraych, who had moved to San Francisco with the company. "On Friday nights there would be no cars on the map."

After initially financing the company himself, in late 2009, Wolpert set out to raise capital. Three groups of Bay Area angel investors agreed to put in a total of less than a million dollars. That was another mistake; it wouldn't be enough. Wolpert was being too careful. "We were bringing a knife to a gunfight," he says.

Soon after, a golden opportunity beckoned, and Cabulous, like Taxi Magic, missed it. As Wolpert was wrapping up the financing, his phone rang. Wolpert didn't immediately recognize the Texas-tinged voice on the other end of the line. Fresh off his disappointment with Taxi Magic and still searching for a prime seat in the coming transportation revolution, venture capitalist Bill Gurley had heard Cabulous was raising money and asked how much room was left in the round.

Wolpert was surprised to get the call and gave an honest answer—the round was basically full. He mentioned a small number. Gurley said that wasn't enough to interest Benchmark Capital. Gurley then offered some free advice about focusing, suggesting launching first in specific neighborhoods rather than an entire city, before hanging up.

Wolpert probably should have taken Gurley's offer, he mused years later, but it would have required ditching the investors that had already committed. In other words, he would have had to ruthlessly prioritize the best possible outcome for his company, despite his prior personal commitments.

"I was a Boy Scout. I was going to go with the date that brought me," he says.

When UberCab introduced its black-car service in San Francisco in June 2010, Cabulous was the closest thing it had to a crosstown rival. Tal Flanchraych recalls seeing an Uber job posting on Craigslist that spring. It was headlined "UberCab Sr Engineer: ground floor gig at BALLER location startup" and announced that Uber was looking to find engineers to help it build a ground transportation app that was "similar to Cabulous."

Ryan Graves, Uber's first CEO, also reached out to Wolpert, and they met for coffee at the Delancey Street Restaurant on the Embarcadero. Wolpert was friendly but dismissive of Uber's approach.

"We didn't spot the idea that there was this ready group of limo drivers who were sitting in parking lots and at the airport, waiting for a dispatcher to call a limo," he says.

Wolpert's conversation with Graves was pleasant until Travis Kalanick joined them at the restaurant. He asked bluntly: "Are you going into limos?" Wolpert said he wasn't. He believed that was a bad idea and knew that the yellow-cab drivers using Cabulous would get upset if the company started aiding their more lightly regulated competition. All those drivers knew where Wolpert lived; he had been personally answering their technical questions from his apartment near the ballpark. "We've made our bet," he assured them. Graves and Kalanick left quickly.

Over the next few months Cabulous warily circled Taxi Magic, drawing up expansion plans and jockeying to sign up taxi fleets. Then UberCab, with its more elegant app and deluxe black-car experience, started gathering momentum, plaudits, and venture capital, and it eventually blew both companies out of the water.

When he heard that San Francisco officials had served UberCab with that first cease-and-desist notice, adorned with a head shot of Ryan Graves, Wolpert believed it was justified. Regulation had a purpose. Taxi prices needed to be strictly controlled so that grandmothers could afford a ride home from the supermarket. He knew the pricier black cars were regulated more lightly, but by law they had to be summoned in advance, limiting their ability to compete with cabs.

Wolpert felt that UberCab blew up that distinction with new technology, allowing riders to electronically summon a town car with little forethought, just as they would hail a yellow cab on the street.

What he found even more objectionable was that Uber was using the iPhone as a taximeter to calculate fares. Meters are traditionally calibrated and closely monitored by cities' weights and measures departments to protect customers from price gouging. He and Graves, who had been friendly since their first meeting and had attended some MTA meetings together, argued about it over the phone. "Hey, let's completely disregard decades of regulation!" Wolpert shouted at him. "How is this a good idea?"

"I guess we don't have anything left to talk about," Graves said, and hung up. They never spoke again.

Wolpert left Cabulous in 2011 after Uber was starting to drive laps around it, ceding the fight to a more experienced CEO. Years later, when the company changed its name to Flywheel, the brand started appearing all over the exteriors of San Francisco taxis owned by DeSoto Cab, part of a co-marketing agreement.[2] Wolpert now gets emotional when he sees it. "I may not be rich from this, but I changed the face of the city," he says. "It makes me happy."

He concedes that it was probably a mistake to partner with yellow-cab drivers and taxi fleets, which were handcuffed by regulation and ill prepared to counter the disruptive threat posed by Uber. "It was like watching a shark devour a seal," he says. "We are living in an era of robber barons. If you have enough money and can make the right phone call, you can disregard whatever rules are in place and then use that as a way of getting PR. And you can win."

He is now back at IBM. At the end of our talk, on his way out the door, he is overcome by the same concerns that gripped DePasquale—that his failure to seize an enormous opportunity will be made public.

"Please don't ruin my career," he says.

Bad decisions and imperfect technology weren't the sole province of the car-service startups that preceded Uber.

An online home-sharing service called Couchsurfing won devotees and attention five years before the rise of Airbnb. It wasn't bad timing, stubbornness, or chronic niceness that doomed it, but something just as deadly in the cutthroat world of business: idealism.

Couchsurfing was the inspiration of a young, broke programmer from New England named Casey Fenton. Fenton's vision was nearly identical to the one that was later articulated by Brian Chesky and Joe Gebbia, right down to Couchsurfing's weighty mission statement ("to connect people and facilitate inspiring experiences"). Even the brand names Airbnb and Couchsurfing were similar, each implying an uncomfortable night's sleep that could leave one sore in the morning.

Unlike Chesky, Fenton grew up with divorced parents and spent his early years shuttling between homes in New Hampshire and Maine. He was the oldest of five kids in a family that was poor and periodically on food stamps, and he left home as soon as he could, graduating high school early with a determination to see the world and have "an interesting life."

During his college years in the late 1990s, Fenton bought random plane tickets to different parts of the world and depended on the kindness of locals to take him in. He traveled to Cairo and ended up surreptitiously climbing an ancient pyramid with a taxi driver; later he went to Iceland and spammed a large selection of the student directory at the University of Iceland, unabashedly asking for a couch to crash on because he couldn't afford a hundred-dollar-a-night bed at the local youth hostel.

Fenton had magical interactions on those trips and wanted to share this type of experience with the world. He registered the web address Couchsurfing.com in the late 1990s, then spent a few years consulting for startups and working in Alaska politics before he got around to launching the site. His partners in the endeavor were Daniel Hoffer, a Harvard graduate and travel aficionado turned entrepreneur who had previously hired him to do programming work, and two of their friends.

Couchsurfing opened in 2004 and drew a young, itinerant crowd that was less interested in accumulating wealth than sharing it. Like Airbnb years later, hosts and guests wrote their own profiles on the site and reviewed each other after a stay. One ingenious element of the service was how the company confirmed people's identity in the years before people could use their Facebook profiles.

Couchsurfing asked for a user's credit card and sent a postcard containing a verification code to the address associated with the card. When the user typed the code into the site, he or she was verified. The company charged twenty-five dollars for the service, and for years it was its only stream of revenue.

Fenton infused his company with romantic notions. He pitched it not as a marketplace for accommodations but as a way for travelers to meet new people, have novel experiences, and make the world a warmer, more inviting place. Earnestly practicing what he preached, he registered the company as a nonprofit in the state of New Hampshire. Years later, sipping organic lentil soup at the Plant Café Organic in San Francisco, he conceded there had been an awful lot of naïveté involved at the time. "This is what you do when you don't understand corporate entities," he says.

Because of its nonprofit status, Couchsurfing didn't have employees or a real office. Instead, it had hundreds of nomadic volunteers who roamed the world using the site and slept on one another's couches. The four founders worked from their disparate homes, with Fenton crashing with Hoffer in Palo Alto for months at a time. Occasionally the most active members of the community would live together for a few months in places such as Thailand, New Zealand, and Costa Rica, churning out improvements to the site.

By 2008, the company had a few dozen paid employees and over two thousand volunteers, all in different time zones and frequently on the move. Not surprisingly, the site was ugly, outdated, and difficult to use. Then AirBed & Breakfast emerged.

New Hampshire had just informed the company that it was improperly registered and had to change its tax-exemption status. The founders could not agree on what to do. By now Daniel Hoffer was a business-school graduate and a product manager at the Silicon Valley security firm Symantec. He lobbied his co-founders to shift to a for-profit model and start charging guests per stay to generate some real revenue.

Fenton was adamantly against this, arguing that hosts and

guests exchanging money would dilute the purity of the experience. Instead, he initiated a prolonged and expensive effort to convert that nonprofit status in New Hampshire into a federal 501(c)(3), a tax-exempt nonprofit organization.

"Life is short," Fenton said in an interview at the time, explaining his thinking. "I want to do meaningful things. Money seems like it can come easily. If you just want to focus on money that's fine. I want to do some things that I find more interesting."[3]

Hoffer suspected that Airbnb was a threat to Couchsurfing. In 2008, before they entered Y Combinator, when Chesky and Joe Gebbia were still flopping around San Francisco and soliciting everyone they could find for advice, the Marine turned venture capitalist Paige Craig introduced them to Hoffer and the three of them met for pizza one night in the city's Mission District.

Chesky and Gebbia peppered Hoffer with questions about Couchsurfing and the challenge of building trust between total strangers who are sleeping under the same roof. The dinner was friendly, but Hoffer sensed there was trouble ahead. "They were clearly approaching it intelligently and they seemed smart. I was very threatened by them," he says.

Chesky later told me that he was not impressed by Couchsurfing. "I had done enough product development to know that there could be fifty companies that make chairs but it doesn't matter. The one who wins is the one who makes the best one." Couchsurfing, he said, was like an amateurishly made chair; it had a chaotic design, no sense of hospitality, and no payment mechanism. "To me it was a totally different thing," he says. Comparing them "is like saying every piece of furniture is the same."

After the dinner, Hoffer called Fenton and the other co-founders and begged them to surrender the bid for nonprofit status. They refused. They were all tired of having the same argument over and over.

Years later Hoffer could only reflect on a very substantial what-if. Chesky and Gebbia wanted a mentor and collaborator; pursuing that relationship could have resulted in getting in on the ground floor of an enormous opportunity. "I prioritized my loyalty to Casey and the other founders and to the Couchsurfing community," Hoffer says, speaking slowly in the conference room of the venture capital firm where he now works. "So it was...a choice. Which probably cost me a billion dollars."

The rest of the Couchsurfing story is not a pretty one. Hoffer replaced Fenton as the CEO of Couchsurfing in 2010. The IRS rejected the 501(c)(3) application on the basis of, well, common sense: the company was saving users money on travel lodgings, not necessarily facilitating an exchange in cultural values or making the world a better place.

Couchsurfing suddenly had to raise capital to finance the expensive shift to for-profit status and to pay its back taxes. It raised $7.6 million from a group of investors led by — who else? — Benchmark Capital, which thought it saw an opportunity to compete with Airbnb in the suddenly fashionable home-sharing category.

Benchmark's partner on the deal, former Facebook executive Matt Cohler, must have realized that he had placed a bad bet. After the conversion to for-profit status was complete, he fired Hoffer, Fenton, most of the employees, and all of the volunteers. Couchsurfing's most rabid users unleashed a torrent of vitriol on various online bulletin boards, and the site was overtaken by Airbnb in popularity. The new CEO of Couchsurfing lasted less than two years.

There remains one big tale left to tell in this account of entrepreneurial also-rans. It's the story of a company called Zimride.

Just as eBay let sellers hawk unused items from their attics and the stuck-in-time Craigslist allowed people to sell old cars, used futons, or even their spare time to do odd jobs, the founders of Zimride realized the same principle could be applied to the empty seats of cars on long road trips. Zimride never truly captured mainstream attention. But the company would end up playing a significant role in the coming battle royal between the upstarts of Silicon Valley and around the world.

The tale starts with Logan Green, a young, introverted software engineer who grew up amid the transportation chaos that was Los Angeles in the 1990s. In high school, Green got a part-time job working for the celebrated video game entrepreneur Nolan Bushnell, the founder of Atari and one of the first bosses of Apple co-founder Steve Jobs. Green went to the hippie-ish New Roads High School in Santa Monica and for years navigated the gridlocked city streets in his beat-up 1989 Volvo 740, commuting to Bushnell's gaming company, uWink, in Playa del Rey.

The drive was only six miles but could take more than half an hour. "I just recall having this feeling of seeing everyone stuck in traffic," he told me years later. "There were thousands of people heading in the same direction, one person in each car. I thought, *If we can just get two people in the car, you could get half these cars off the road*."

Green was so disgusted with Southern California traffic that he left his beat-up Volvo at home when he enrolled at the University of California at Santa Barbara, committing himself to public transportation. "I wanted to push myself and to see what it was like getting around," he said. During his sophomore year in 2002, he learned about the East Coast car-sharing club Zipcar, which allowed members to take out vehicles for flexible periods without owning them.

After failing to get Zipcar to put cars in Santa Barbara, Green started a car-sharing program at his school. The university purchased a small fleet of Toyota Priuses and Green devised a system so that students could book the cars on a website and unlock doors with special radio ID cards and access codes.[4] He spent two years on the project and a couple thousand students started using it.

But getting home to Los Angeles for holidays and to visit his girlfriend (later his wife), Eva, was always an adventure. On long-distance bus lines he would meet recently released inmates from the county jail, their possessions stuffed into trash bags. He also experimented with Craigslist, which had a ridesharing channel before the term *ridesharing* was widely popular. Though his experiences with long-haul ridesharing were generally good, Green was never totally comfortable. Climbing into a car with a stranger was unnerving.

Buoyed by these experiments, Green became the youngest member of the Santa Barbara Transit Board and got a full education in the grim economics and politics of the public bus system. Seventy percent of every bus ride had to be subsidized by the city. The quality of service was low but attempts to raise fares and add sales tax frequently buckled under a wall of local opposition.

Over the summer of 2005, Green and his best friend in high school, Matt Van Horn, decided to travel overseas. They were set to go to Cuba, which was illegal in the United States at the time, until Van Horn's mother started worrying and bribed her son to go to Africa instead by paying for part of his plane ticket.

In Africa, Green and Van Horn went on a monthlong tour that took them from South Africa through Namibia and Botswana, and to Zimbabwe. It must have been entrepreneurial fate, because the young men were thunderstruck by what they saw in Victoria Falls. Zimbabwe was exceedingly poor and few people owned a car. Instead, people piled into minivans driven by unlicensed taxi drivers. "It wasn't that well organized but there was such efficiency," recalls Van Horn. "It didn't make sense to drive a car unless every seat was filled and everyone was paying a little bit of gas money."

During Logan Green's senior year back in Santa Barbara in the fall of 2005, the pieces started coming together in his mind—the ridesharing channel on Craigslist; the crowded vans in Victoria Falls; the intractable flaws of the public-transit systems. He started working on a concept he called Zimrides (short for Zimbabwe rides). The idea was to use the internet to fill every open seat in every car.

It seemed like perfect timing. That year, the rising social network Facebook started letting other internet companies introduce services that incorporated people's membership profiles. This was the missing element in services like Couchsurfing. By seeing potential riders' real names, photographs, and social connections, people would be more comfortable sharing their cars with them. In December of 2006, Zimride's first app, called Carpool, allowed university students to post on Facebook, specifying where they were traveling to and looking for rides with others going in the same direction.

Across the country, a recent graduate of Cornell saw the app and was fascinated. A student at Cornell University's school of hotel administration, John Zimmer had learned that the key to running a profitable hotel business was high occupancy and great hospitality. The transportation status quo offered neither. "If you take public transit and taxi and think of that as a hotel, those would be hotels you would not want to stay in," he told me later. "They would be failing businesses." Inspired by Zimride's Carpool app, Zimmer (the similarity of the company's name to his own was coincidental) got a friend to introduce him to Green and they decided to team up in a virtual cross-country partnership.

The pair worked on the project part-time with Van Horn, who had moved to Arizona to attend law school, and another developer. Naturally, it was a side project. They introduced the app at Cornell, where students quickly embraced it. They also found it randomly adopted in places like the University of Wisconsin–La Crosse, where over the holidays students would migrate en masse to nearby cities such as Madison (a two-hour drive) and Chicago (a four-hour drive). Eventually the founders started pitching the service directly to schools. Universities could pay a few thousand dollars a year and get their own specialized version of it.

Enjoying a little momentum, Green and Zimmer tried raising money up in Silicon Valley. No one would meet with them. Then, out of the blue, Green got an e-mail from an eBay executive and angel investor named Sean Aggarwal, asking to invest. Green thought it was a scam and asked Van Horn to accompany him to a public location, Coco Chicken in Fremont, California, to see if

Aggarwal was a real person. He was, and he wanted to write a check. They spoke for a few hours that day and Aggarwal became the company's first investor and adviser.

Now they had a little bit of money and guidance. With the new cash, John Zimmer randomly purchased a frog costume and a beaver costume. The founders would wear them while handing out Zimride flyers at colleges.

Over the summer of 2008, Zimmer and Green moved to Palo Alto and lived together in a place not far from Facebook's offices.[5] They were strangers in a strange land, roommates and office mates in a cramped two-bedroom apartment that abutted the backyard of future Yahoo CEO Marissa Mayer. At night, sitting alone, they would hear Mayer's loud outdoor parties and awards ceremonies. When the names of people were called out, they would rush to Google and type in the names to see who they were.

When they weren't eavesdropping on the neighbors, they were watching Zimmer's former employer Lehman Brothers go bust during the market crash. With the economy cratering, they reasoned, carpooling might suddenly come into vogue. "We were sitting there thinking this is going to be great for business," Green says, "but this is going to be horrible for getting the company funded."

The company grew nicely for another year, then caught the eye of a partner at Floodgate, the venture capital firm that had missed out on Airbnb when the website crashed during Brian Chesky's presentation. By then, Floodgate had realized its mistake, and the Zimride duo, aware of Airbnb's rising reputation, had enough sense to reference the home-sharing startup in their pitch deck. One of the firm's partners, Ann Miura-Ko, who was attracted by the pair's passion for the economic and environmental benefits of ridesharing as well as by their steely determination in the face of what had already been a long slog to launch the company, led a $1.2 million round of financing.

"You want to back entrepreneurs who, even when the chips are down and things aren't working and everyone says this isn't meant to be, have so much love for the idea and so much passion that they just persevere," says her partner Mike Maples Jr. "Startups are very romanticized and most people are completely clueless about how you just have to will it into existence."

Despite that new capital, Zimride still sputtered. The founders pitched the carpooling service to new universities and to some companies, such as Walmart—take Zimride every day to work!—and then opened the website to regular folks. The startup ran buses between major cities, like Los Angeles and San Francisco, and from cities to the Coachella and Bonnaroo music festivals. Sometimes Zimmer and Green would even drive. They raised another six million dollars in funding from venture capitalists in 2011 and moved up to San Francisco, to the fashionable South of Market District, where an uprising of startups was starting to shift Silicon Valley's center of gravity north.

But when they were being honest with each other, Green and Zimmer had to admit that Zimride wasn't going to get big enough to change the world. Internet marketplaces thrive when buyers and sellers are matched in ways that wouldn't otherwise be possible, saving everyone time and money. Even the most enthusiastic carpooler used Zimride only a few times a year. And the service was helping them find fellow travelers, essentially replacing Craigslist and the old cork bulletin boards in the university quad, but not doing much else. "It was a big vision but it wasn't the right execution," Green says.[6]

All the lethal mistakes of the other nonstarters were wrapped up in Zimride. The founders were too nice. They were idealistic. Their idea was too early—the great wave of smartphone ubiquity and social networking was just gathering momentum. But they were also pragmatic, and they believed in that Silicon Valley notion referred to as "the pivot." As long as there is money in the bank, it's never too late to change business models and seek more profitable pastures.

In early 2012, the founders and their engineers met frequently over a three-week period to discuss what to do next. Impressed by the success of Uber's black-car business, they got excited about a mobile version of Zimride that would let regular people share their vehicles not on long trips or daily commutes but every day, anytime, from one point to another within big cities. Inspired by a giant orange felt mustache that decorated the cubicle of an employee, John Zimmer decided to give every driver a pink mustache to put on his or her car fender; the mustache would make the car stand out and appear friendlier to those who might be wary of climbing into a car with a stranger.

At first they referred to the new service as Zimride Instant, then changed the name to something a little catchier: Lyft.

But now we're getting ahead of ourselves.

CHAPTER 4

THE GROWTH HACKER

How Airbnb Took Off

Son, no one from the internet is going to pay you a thousand dollars.

—Paul Blecharczyk to his son Nathan

Greg McAdoo knew all about the nonstarters. A year and a half before he met the founders of AirBed & Breakfast, the New York City–born venture capitalist had an epiphany about how to consolidate and streamline the vacation-rental market. Small-business owners in the travel industry, like proprietors of bed-and-breakfasts, usually had only enough money to advertise locally. The internet could let them reach travelers around the world.

Exploring this thesis, he visited more than half a dozen web outfits, such as LeisureLink and Escapia, and started watching HomeAway, an Austin, Texas, company that was gobbling up rivals like the VRBO—Vacation Rentals by Owner—in an effort to create a dominant network of vacation properties. McAdoo spent nearly a year sizing up these companies, but he wasn't convinced that any of them had a particularly novel approach. "It was a very fragmented market and it was never clear how it should be presented online," he said many years later. "Frankly, I had moved on."

Then, in early 2009, he sat down for coffee with Y Combinator chief Paul Graham and started talking about the mental toughness that founders needed, and Graham pointed across the room to the Airbnb guys as prime examples.

McAdoo introduced himself to Brian Chesky, Joe Gebbia, and Nathan Blecharczyk that day and was struck by their approach. The vacation-rental startup founders he had talked to were trying to make the experience better for travelers; the Airbnb guys wanted to make it better for *hosts*. It was the first of many meetings McAdoo and his partners at Sequoia Capital would have with the Airbnb founders over the next few months and one way in which YC, as the high-profile startup school is known in Silicon Valley, would radically change the prospects of the struggling company.

Airbnb had only narrowly gained admission to that winter's YC program, thanks in part to its unlikely cereal gambit. After the founders got in, Blecharczyk apologetically said a temporary good-bye to his fiancée in Boston, moved back to the Rausch Street apartment, and installed himself on the living-room sofa. The founders drove forty-five minutes to get to YC's offices in Mountain View, on a street optimistically called Pioneer Way, where they typically set up shop on the long trestle tables in the main dining hall.

At YC, Chesky, Gebbia, and Blecharczyk had constant access to Graham, the closest thing Silicon Valley had to Yoda from *Star Wars*. Since selling his e-commerce company, Viaweb, to Yahoo during the first dot-com boom, PG, as he was called, had become a font of startup aphorisms, such as "It's better to have a hundred people that love you than a million people that sort of like you," and "Don't worry about competitors; startups usually die of suicide, not homicide." He was in his early forties and typically wore the I-don't-care-about-your-social-customs sartorial combination of cargo shorts, a polo shirt, and sandals.

At the time, the global economy was cratering, unemployment was skyrocketing, and Graham's advice was more sobering than usual. At one point he warned Airbnb and the other fifteen startups in the program that investors were spooked, so they should all make sure they had a graph in their presentations with a single line racing up and to the right, demonstrating increasing profits. The Airbnb founders, who hardly had any revenue, let alone growing profits, mocked up that hypothetical chart and taped it onto the bathroom mirror at Rausch Street.

Despite the economic meltdown, the founders were determined to make the most of their turn of fortune. They stuck around each night, peppering Graham and his staff with questions. "We were the students that were just relentless," Chesky says.[1]

Graham, still skeptical of the home-sharing concept, asked a blunt question: Was the site working anywhere? New York City, the founders replied, where around forty people were making rooms in their homes available for short-term rental. "Well, what are you sitting around here for? Go out there and talk to these people," Graham said.

While Blecharczyk stayed behind to code, Gebbia and Chesky flew to New York over a long weekend and started meeting hosts. One obvious problem was that hosts weren't presenting their properties online in an appealing way—the photos were grainy and usually taken with the primitive cell phones of the time. They reported this observation back in Mountain View, and Graham compared it to a challenge he had encountered at the online marketplace Viaweb, where he had to show naive retailers how to sell on the internet. "What they needed to do was teach their hosts how to sell," Graham says. "That was the missing ingredient."

So in what has become a bit of oft-repeated Airbnb lore, Chesky and Gebbia returned to New York regularly over the weekends that winter after e-mailing hosts that the site was sending a professional photographer to their homes for free.

Once in the city, they rented a high-quality camera and trudged around in the snow, knocking on doors and taking pictures of people's bedrooms and backyards. "We were on a budget. I remember deliberating every little expense, like the quality of the tripod and whether we should go for the nice one or not," Gebbia says.

In the parlance of the Valley, this kind of activity did not "scale." It was a wildly inefficient use of their time. But it helped the founders tune in to the needs of their earliest users and to recognize that large, rich, colorful photos of homes and good profile pictures of the hosts would make the experience on the site more compelling. "Paul was the first person to give us permission to say, It's okay to think about things that may not scale, to break away from the mythology of Silicon Valley," Gebbia says. "We could actually think creatively around how to grow the business."

Gebbia and Chesky logged a lot of miles that winter of 2008–2009. They spent most weekends in New York and on Tuesday mornings flew back to San Francisco; Blecharczyk would pick them up at the airport, and the three of them would race down to Mountain View to make it in time for the weekly YC dinner. "They were never late for anything and were the first to show up and the last to leave," Graham says. He was starting to believe—first in the dedication of these entrepreneurs and then, slowly, in the concept itself. "How are the airbeds doing today?" was his standard greeting to them.

But Graham was still having trouble wrapping his mind around the idea of people actually sleeping on airbeds. Finally he identified the real opportunity as "eBay, but for spaces" and urged them to think of their brand as comparable to the auction giant's. By the end of the program, the founders had moved their website from Airbedandbreakfast.com to the abbreviated Airbnb.com.

Greg McAdoo was also getting excited about the concept. Nobody else in the vacation-rental market had taken the time to visit the actual hosts and survey their needs, and no one else had the three men's facility with the emerging tools of social media, like community meet-ups, online ratings, and Twitter.

Moving quickly in advance of demo day, when rival investors would get a peek at Airbnb, McAdoo introduced a few of his Sequoia colleagues to the company. There were some prescient questions.

"Have you thought about the legality of this?" asked Mark Kvamme, a longtime Sequoia partner.

McAdoo believed that it was too early to judge how such a novel activity would fit into existing laws governing the hotel industry. "These businesses either work or don't work based on whether or not they are good for consumers," he recalls telling Kvamme.

McAdoo returned to the YC offices the day before demo day and sealed the deal with the founders in a side room, convincing them not to appear onstage to deliver their already scripted presentation. Sequoia, one of the marquee firms of Silicon Valley, invested $585,000 for approximately 20 percent ownership of this small, unproven startup that had struggled for much of the past year and a half. Along with its participation in later rounds of fund-raising, Airbnb would turn into Sequoia's most profitable investment ever, exceeding even the returns on its home-run bets Google and the chat service WhatsApp. The December 2016 value of its stake: $4.5 billion.

But back then, in March of 2009, that success was still far off. Once Airbnb graduated from YC, more than a year after they had started the company, the founders returned to living and working on Rausch Street, facing many of the same challenges of the previous year: there was little variety on the site in most tourist destinations and an anemic rate of revenue and listings growth.

The startup hadn't solved the tricky chicken-and-egg problem that confronts the creators of online marketplaces. The relatively few listings on the site drew few guests looking for travel accommodations, and the paucity of guests didn't inspire potential new hosts to embrace the unorthodox concept of making their homes available to total strangers over the internet.

The Airbnb founders like to talk about some of their more ham-handed attempts to ignite their marketplace that first year, but none of them account for how the company actually achieved liftoff. For example, while Blecharczyk stayed behind to code, Chesky and Gebbia continued to try to build up early listings by visiting New York, Las Vegas, and Miami, among other cities, and organizing meet-ups with any hosts they could find.

During one meeting McAdoo suggested another way to boost growth: urging property-management companies that controlled multiple listings to add them to the site. Chesky hired three sales interns that summer to cold-call such firms. Then, in the fall of 2009, he and Gebbia visited Europe. In Paris they stayed in the spare room of a native Parisian who was charming and hospitable. Chesky remembers it as a magical trip. The following week they went to London and stayed at a home that had been listed by one of the property-management firms. There was no host present and the two found it to be a hollow experience. "It didn't have the love and care and it felt like it wasn't in the spirit of what Airbnb was," Chesky says.

He returned to San Francisco and says he ended the cold-calling operation. Whether Airbnb was actually committed to keeping managers of multiple properties off the site would later become a topic of blistering controversy. Such opportunists would flock to the site anyway, and cities were forced to consider how to deal with them and whether to regulate Airbnb like conventional hotels.

The founders seemed to move slowly on everything. McAdoo remembers them as being a little too "wonderfully frugal," reluctant to spend their new venture capital, which is ironic considering their later profligacy on elaborate corporate offices around the world. "On the one hand, that is fabulous," McAdoo told them, discussing the high bank balance. "On the other hand, guys, we need to invest in the business." They also moved glacially in signing on new employees, declining at first to even hire customer-support help. (The only phone number on the site redirected the caller to the personal cell phone of Joe Gebbia.) The founders spent six months looking for the first full-time engineer before they finally settled on Nick Grandy, a fellow Y Combinator alumnus who had abandoned his own startup.

Grandy, who left Airbnb in 2012, remembers working out of a cluster of desks in the apartment living room, at first with the distracting chatter of the sales interns on the phone cold-calling hosts. One early challenge, he remembers, was getting hosts to actually reply to guests' messages on the service. The solution was to make a user's response rate—"the host responds to 75 percent of messages," for example—visible on the site.

The founders worked seven days a week but there was a spirit of camaraderie and plenty of goofy fun. They would break occasionally to go to the gym or hang out on the roof. Once a week, they went to a nearby park on Folsom Street for "recess," to play kickball or even a game of tag. On Friday they usually went to a bar for happy hour.

Eventually ten employees worked out of the crowded apartment. Chesky had to interview job candidates in the stairwell for privacy; employees took important calls in the bathroom.[2] The bedrooms became offices; Gebbia slept on a mattress on the floor until he finally rented another unit in the building, and Chesky started living entirely out of a suitcase in rooms he rented around town on Airbnb. During one month, he stayed in the captain's quarters of a Norwegian icebreaker that was docked in the Bay Area.[3] And he left behind his Honda Civic and commuted to the apartment by taking Uber, the suddenly fashionable black-car service that was sweeping San Francisco. The Airbnb employees marveled at the magic and simplicity of the Uber app, says Grandy, and it inspired the team when they developed the first Airbnb app for the iPhone in 2010.

Chesky was moving slowly, but at the same time, he was frustrated that his imagined success wasn't arriving quickly enough. "Every day I was working on it and thinking, *Why isn't it happening faster?*" he told me.[4] "When you're starting a company it never goes at the pace you want or the pace you expect. You imagine everything to be linear, 'I'm going to do this, then this is going to happen and this is going to happen.' You're imagining steps and they're progressive. You start, you build it, and you think everyone's going to care. But no one cares, not even your friends."

To understand the spark that finally ignited the Airbnb inferno, it's imperative to explore the background of Nathan Blecharczyk, the tall, seemingly unflappable engineer of the group, the co-founder who always stayed behind while his partners traveled the world.

Blecharczyk was twenty-four at this point but was already a technical wizard. He had coded the entire site himself, using what was then a new open-source programming language called Ruby on Rails. He devised a flexible, global payment system that allowed Airbnb to collect fees from guests and then remit them to hosts, minus the company's commission, using a variety of online services such as PayPal. He had also presciently hosted the site on the nascent Amazon Web Services, a division of the e-commerce giant that allowed companies to rent remote Amazon servers via the internet only when they needed them, a huge cost savings and efficiency advantage that would power an entire wave of new businesses.

"Joe and I would have crazy dreams and visions," says Chesky of his co-founder. "Nate would find a way, without compromising the vision, to make the wildly impractical possible."

But that wasn't the full extent of Blecharczyk's talents.

Nathan Underwood Blecharczyk was born in Boston, the son of a homemaker mom and an electrical-engineer dad who worked for a local company that made industrial equipment. Blecharczyk's father, Paul, taught him and his younger brother to be curious and to question how things worked. He would have his sons do mechanical tasks around the house, and he would bring home discarded equipment, like an old Xerox copier machine, and invite them to take it apart in the backyard. "There is no job too big or too small for PB and sons," he would say to his boys.

Soon, young Nate was consumed with computers. According to

family lore, home sick one day from middle school at age twelve, he took a book about computer languages off his dad's shelf and devoured it. For Christmas, he asked for a book about Microsoft's programming language QBasic, and he plowed through that one in three weeks.

Blecharczyk ran cross-country at his Boston public high school and excelled in his classes, but at home he had a far less conventional life. After learning to code, he started writing increasingly sophisticated computer programs and giving them away on the internet, asking for voluntary donations. An early such shareware program allowed computer users to place digital sticky notes on their screens. Later on, another program of his interfaced with America Online, the dominant online network of the time, which was then walled off from the broader web, and gave programmers a way to send internet messages into the e-mail and IM accounts of AOL members.

Soon after he posted that program, Blecharczyk got a phone call from someone who had seen it. The person offered him a thousand dollars to write a similar e-mail tool. When he told his dad about the offer, Paul Blecharczyk responded: "Son, no one from the internet is going to pay you a thousand dollars."

Nevertheless, Blecharczyk wrote the program and earned the money. He later found out his customer had himself been hired to create it and was merely subcontracting out the work (and was probably paid more than a grand). The customer then introduced Nathan to his client and to other potential clients, and suddenly Blecharczyk was earning considerable money writing a variety of tools for a nascent industry. Its practitioners innocuously dubbed it "e-mail marketing." The world, of course, came to know it as something else: spam.

Throughout high school and into college, Blecharczyk wrote customized tools for spammers. He eventually developed a suite of e-mail marketing products to help them organize and orchestrate their campaigns and maneuver around internet service providers that were desperately attempting to shut off the spam deluge. The orders poured in, as did the cash. He called his company several names at various times, including Data Miners and, eventually, Global Leads, which he incorporated in the State of Massachusetts after his freshman year at Harvard in 2002. At first he couldn't accept credit cards, Blecharczyk recalls, so he had spammers enter their bank account details on his site, and then he printed the bank numbers on blank Office Max checks, wrote down the amounts he was due—typically around a thousand dollars—and went to the bank to deposit them. "Amazingly, this is legal," he says, recounting his early success with delight. "I was literally printing money!"

At the end of every week and after every three months, he gave his parents a financial report. Naturally, Paul and Sheila Blecharczyk were mystified. "This was a whole new world," Blecharczyk says. "The internet had just been born. I don't think anyone really knew what to expect or what this was."

The spam operation earned Blecharczyk close to a million dollars, he says, and paid his college tuition at Harvard University and more. It also earned him a spot on an online blacklist called Register of Known Spam Operators, maintained by an anti-spam London-based organization called the Spamhaus Project. On its page devoted to Data Miners, Spamhaus alleged that Blecharczyk often used the names Nathan Underwood and Robert Boxfield and appeared to have set up a service that offered spammers access to a range of accounts, called relays, outside the United States that would disguise and anonymize their e-mail campaigns. "Data Miners (aka: Nathan Underwood Blecharczyk) is one of the main sources of broken/open e-mail relays (used by spammers), and the tools to help locate and exploit them,"[5] Spamhaus reported.

Blecharczyk says he shut his business down in 2002 to focus on his college studies because the work was taking up all his time. A Harvard classmate later recalled Blecharczyk telling him that he had received threatening letters from the Federal Trade Commission about his activities.[6] (Blecharczyk does not recall this.)

He discusses all this years later from Airbnb's offices and is unapologetic about how he earned his first considerable fortune. "All this was new," he says. "There were frankly no rules around it." That is technically true—the Federal CAN-SPAM Act that made sending or facilitating spam a federal crime was not passed until 2003. But for years before that, spam was a well-known scourge that frustrated e-mail users and overwhelmed internet companies.

"It's part of being a pioneer," he says. "It's not just exciting to build things but to explore new fields and to recognize what comes with that is a lot of uncertainty. That's very true today and it has been true of Airbnb. It's a whole new concept, around which there haven't been many rules."

When Nathan Blecharczyk graduated from college, he was not just a skilled programmer but the embodiment of a new Silicon Valley hero: the growth hacker. Growth hackers use their engineering chops to find clever, often controversial ways to improve the popularity of their products and services. Blecharczyk, it turned out, was an exceedingly good one.

That makes the mysterious rise of Airbnb in the year after its graduation from Y Combinator easier to understand. Two other apartment-listing services were far larger: Couchsurfing, which was still laboring under the disastrous effects of its nonprofit status, and Craigslist, the popular and practical online bulletin board that hadn't changed much in thirteen years. Craigslist had a huge audience; in 2009, it had forty-four million unique visitors a month in the United States alone,[7] with active apartment rental and home-sharing channels in many of its 570 cities.

Recognizing this fact, Airbnb designed two clever, somewhat devious schemes to usurp Craigslist's advantage. While Airbnb has always minimized the impact of these programs, each bore the unmistakable mark of Nathan Blecharczyk.

In late 2009, a few months after it had graduated from YC, Airbnb appeared to create a mechanism that automatically sent an e-mail to anyone who posted a property for rent on Craigslist, even if that person had specified that he did not want to receive unsolicited messages. If the apartment was listed in, say, Santa Barbara, the e-mail would read: "Hey, I am e-mailing because you have one of the nicest listings on Craigslist in Santa Barbara and I want to recommend you feature it on one of the largest Santa Barbara housing sites on the Web, Airbnb. The site already has 3,000,000 page views a month." All these e-mails were identical except for the city, and they typically emanated from a Gmail account bearing a female name.

Dave Gooden, another online real estate entrepreneur, recognized the soaring popularity of Airbnb in 2010 and became curious about it. Suspecting what was going on, he posted a few dummy listings on Craigslist and then wrote a blog post in May 2011 about his findings, concluding that Airbnb had registered Gmail accounts en masse and set up a system to spam everyone who posted on Craigslist. He described Airbnb's activity as a nefarious, "black-hat" operation. "Craigslist is one of the few sites at massive scale that are still easily gamed," he wrote. "When you scale a black hat operation like this you could easily reach tens of thousands of highly targeted people per day."[8]

After Gooden's post, a few technology blogs picked up the story and Airbnb was put on the defensive.[9] Its explanation was that it had hired contractors who were behind the effort to spam Craigslist users. "One of the lessons you learned is you have to be very close, provide constant management and guidance to the people you're working with," Chesky said when I asked him about it onstage at an industry event after Gooden's blog post.

A few years later, Blecharczyk offered a little more detail. They had hired foreign contractors on eLance, an online staffing service, and were paying them per lead, or for every new host that would list on Airbnb. "Many companies bootstrap themselves off of finding a user segment on Craigslist and then building a better experience and going after those users," he says. The whole effort, he insisted, was ineffective because Craigslist users typically weren't looking to rent their rooms to vacationers but to find roommates or longer-term tenants. "It did not end up driving any meaningful business," he says.

But another strategy clearly did. A few months after the bulk-e-mailing campaign to Craigslist users, Airbnb tried a new tactic. Instead of luring Craigslist users to Airbnb, the company did the opposite: it allowed users to take a streamlined version of their elegant listing on Airbnb and then cross-post it with a single click on Craigslist. "Reposting your listing from Airbnb to Craigslist increases your earnings by $500 a month on average," the site informed prospective hosts. "By reposting your listing to Craigslist, you'll get the benefit of more demand, while still being able to use Airbnb to manage and moderate your inquiries."

The tool, which Chesky says was originally the idea of adviser Michael Seibel, was a boon for the company. It established Airbnb as a way to create more visually appealing Craigslist ads and, in effect, dropped ubiquitous Airbnb ads into the network of its largest competitor. "It was a kind of a novel approach," Blecharczyk says. "No other site had that slick an integration. It was quite successful for us."

Other growth hackers noticed this and applauded it as a sophisticated technical achievement. Craigslist has different versions of its site in hundreds of cities, each with different web domains and menu formats. Blecharczyk had designed a way to make it simple for Airbnb to post seamlessly onto the right site. "It's integrated simply and deeply into the product, and is one of the most impressive ad-hoc integrations I've seen in years," wrote Andrew Chen, a fellow growth hacker who would later work at Uber, in an admiring blog post. "Certainly a traditional marketer would not have come up with this, or known it was even possible. Instead it [would] take a marketing-minded engineer to dissect the product and build an integration this smooth."[10]

For years Craigslist did not seem to care about Airbnb's cross-posting tool. The San Francisco company, one of the pioneers of e-commerce, is small, inward-looking, and not particularly growth-oriented, which is why the outward appearance of its website hasn't evolved in more than a decade. (The company did not respond to requests for comment on Airbnb's activities.) Then, in 2012, Craigslist suddenly woke up to this type of activity and sent cease-and-desist letters to several businesses using similar tactics. Chesky says he could not recall whether Craigslist sent such a letter to Airbnb but noted that the cross-posting tool also helped Craigslist because "it made their ads look better. There were people that wouldn't have posted to Craigslist, and Craigslist got new inventory."

Airbnb dutifully removed the tool after Craigslist objected to these tactics, but by then it was too late. Like sucking through a straw, Airbnb was pulling listings and users over from Craigslist. It helped, of course, that its site was better designed and far easier to use and that it was constantly working to provide easier forms of payment, better mobile apps, and a safer experience where hosts and guests used their real identities and reviewed one another.

Blecharczyk also ran productive online ad campaigns during these early years. If people searched Google for an apartment in Boston, for example, Airbnb ads would pop up at the top of the page. Blecharczyk and his marketing team became experts at finding the cheapest and most frequently searched keywords and generating crisp and somewhat pointed ads. "Better than Couchsurfing .com!" some of Airbnb's early search ads blared. Dan Hoffer, the Couchsurfing co-founder, at one point e-mailed Chesky to complain about this technique. He says Chesky apologized, stopped the campaign, and sent him two boxes of Obama O's as a peace offering.

Blecharczyk pioneered a clever use of Facebook's fledgling ad system, which for the first time allowed companies to tailor and target ads to the interests and hobbies that members specified in their profiles. If a user said he liked yoga, for example, he would see an ad from Airbnb on Facebook that announced "Rent Your Room to a Yogi!" If a person liked wine, he'd see "Rent Your Room to a Wine Lover!" and so on.

Facebook ads were cheap and people tended to respond to these eerily targeted messages. There was a whiff of false advertising, of course, because Airbnb was not actually offering a way to rent rooms specifically to yogis or wine lovers. Nevertheless, Blecharczyk says the Facebook ads worked beautifully and powered the company's expansion. Early employees were left marveling at his combination of technical ability and marketing instincts. Says Michael Schaecher, an early marketing employee who joined over the summer of 2010, Nathan Blecharczyk "is one of the best online marketers the world has ever seen."

By that fall of 2010, thanks in large part to Blecharczyk's growth hacks and a sunken global economy that had many travelers looking for online deals, Airbnb was on fire. It boasted seven hundred thousand nights in eight thousand cities booked on the site and introduced a sleek new app for the iPhone, hitching its wagon to the smartphone revolution.[11]

Airbnb was finally looking like a real company, with revenues and a modicum of corporate decorum. Chesky now referred to himself in the media and on the website as the CEO, formalizing a leadership position he had held since the beginning. Gebbia was chief product officer, in charge of defining "the Airbnb experience," according to the company website, while Blecharczyk was chief technology officer. The company even had a new office, a few blocks away from the Rausch Street apartment in a two-story former auto shop on Tenth Street, with a garage door that opened onto the street, awful cell phone reception, and a local homeless population that made good use of the adjoining side street. It had all the charm of a dusty warehouse but it was a real office, with room for new employees.

The founders realized they had to make customer service a

priority. McAdoo suggested they take a few lessons from another Sequoia company, the shoe retailer Zappos, an unconventional e-commerce player that had originally focused solely on shoes and that had won customers' loyalties with free shipping and by accepting returns with no questions asked. When he spoke to the founders a few days later, they had already taken his advice and been to the Zappos headquarters in Las Vegas; they'd toured the company's tchotchke-lined offices, where employees stood in unison to warmly cheer guests, and met CEO Tony Hsieh and his COO, future Airbnb board member Alfred Lin. Amazon had acquired Zappos in June 2009, but its slightly madcap vibe remained intact.

Around this time, Airbnb returned to Sand Hill Road, the seat of the venture capital industry, to raise more money. Blecharczyk's productive Facebook and Google ads were expensive, and Chesky had to keep the coffers full. Seeing the company's growing market opportunity, McAdoo wanted Sequoia to supply the entire round of funding itself, but Chesky had learned at Y Combinator to be wary of giving too much control to venture capitalists, and he insisted on bringing in another firm.

He found a willing investor in Reid Hoffman, the co-founder and chairman of LinkedIn and a partner at Greylock Capital. Hoffman says he was skeptical at first. *Ugh, couch-surfing is not that interesting,* he thought. Then Chesky met him at Greylock's offices on Sand Hill Road over a weekend and spun a compelling vision of Airbnb as the largest hotel chain in the world but one without the expensive burden of maintaining actual buildings or hiring workers like bellhops and maids. "The idea of essentially transforming this massive illiquid asset that existed in most of our lives—the room, an apartment, a house, a unique space—into something that could actually be in an essentially peer-to-peer marketplace is just one of the killer ideas," Hoffman said. "I was like, 'Okay, I'm ready.'"[12]

Hoffman could seize the Airbnb opportunity in part because the other VC firms Chesky approached still didn't get the concept and weren't able to look beyond the obvious risks—that someone could get hurt in an Airbnb, or an apartment could get ransacked, or a host could stash a secret video camera somewhere. They couldn't see a company that might end up appealing not just to twenty-somethings from Europe but to real adults, even retired couples, who were seeking more authentic experiences when they traveled.

Marc Andreessen, the Netscape founder and investor, had just started his own venture capital firm with partner Ben Horowitz when he passed on Airbnb's Series A round of funding. Andreessen liked to say that the goal of their firm, Andreessen Horowitz, was to identify the fifteen or so tech startups every year that actually mattered and back as many of them as possible.[13] The firm took a long look at Airbnb and whiffed. "Marc struggled with the idea that this would be mainstream," Chesky says. Andreessen Horowitz would rectify the oversight the following year and lead the Series B, a less lucrative but still hugely profitable investment.

Another venture capitalist that passed was across Sand Hill Road at a firm called August Capital. Howard Hartenbaum, an investor in the online video-calling service Skype, met with Chesky repeatedly that fall and took the founders to dinner at Alexander's Steakhouse near the new office in San Francisco. Chesky impressed Hartenbaum; he seemed to have poise, intelligence, and a fierce determination to succeed. But Hartenbaum couldn't wrap his head around the numbers. Chesky, emboldened by Airbnb's early momentum, was offering a 6 percent ownership stake in the company for an investment of $4.5 million.

Hartenbaum thought Airbnb could eventually be a two- or three-billion-dollar company. Even in that best-case scenario, it wasn't a big enough ownership stake to influence the outcome of August Capital's half-a-billion-dollar fund and wasn't a significant enough opportunity for Hartenbaum to try to convince a few of his partners, who were skeptical of Airbnb. So he passed. Years later, he was still beating himself up. What he failed to recognize, he said, was that investors would be seized by a wave of euphoria around the upstarts, and that three billion dollars would end up being a radical underestimation of Airbnb's eventual worth. "You can make lots of little type one mistakes all day long," he says. "They are not fatal. This was a type two mistake, which is the one mistake you can't afford to make. Entire funds are made often on one deal. If you pass on it, you are not doing your job as a venture guy."

Despite the fact that August Capital never invested, Chesky would remember that dinner with Hartenbaum well. It was the first time he heard of the people whose names would soon send shudders down his spine: the Samwer brothers.

"This is probably what's going to happen," Hartenbaum told the three founders over steaks that night. "There are these German brothers. If they haven't already, they will soon see that Airbnb is doing very well. They will then raise a ton of money in a very short period of time to create a company that will copy you. Then they will try to get you to buy them. And they will make your life miserable."

CHAPTER 5

BLOOD, SWEAT, AND RAMEN

How Uber Conquered San Francisco

I'm better than I was. I'm more intense. I'm more awesome. The difference is, in the last [startup] I was afraid of failure. Now I'm not afraid anymore. Now I can just have fun and go and kill it.

—Travis Kalanick[1]

Around the same time Brian Chesky was warned about the Samwer brothers, someone much closer to home was making life difficult for Travis Kalanick and the small band of employees at the city's other rising startup, Uber.

The four plainclothes enforcement officers who had served Uber with its first cease-and-desist on October 20, 2010, set off a mad scramble inside the company. Austin Geidt texted photos of the citation to CEO Ryan Graves, who was in the board meeting at First Round Capital. Graves stepped outside to call her, then returned to the office to discuss the situation with Travis Kalanick, Garrett Camp, and investors Chris Sacca and Rob Hayes. The letter threatened penalties of five thousand dollars per ride and ninety days in jail for each day the company remained in operation. But which laws had they broken, exactly? And who in the vast and impenetrable San Francisco bureaucracy was behind the effort to stop a company that was quickly attracting the loyalties of the local tech community?

A few blocks away, on the seventh floor of a former bank building at 1 South Van Ness Avenue, Christiane Hayashi was planning her next move.

As director of Taxis and Accessible Services at the Metropolitan Taxi Agency, Hayashi was the most powerful figure in the city's highly dysfunctional cab industry. She had worked as a deputy city attorney since graduating from UC Hastings College of the Law and had toiled away in environmental law and Y2K compliance. She was no stranger to San Francisco's bare-knuckle politics, where rival democratic factions battled incessantly and corruption often bubbled indiscreetly under the surface. A stint in the Department of Elections, where she managed the transition away from punch-card ballots, was particularly grueling. Hayashi and two other lawyers were accused of mismanaging funds and signing false time sheets. A special counsel eventually investigated and cleared her.[2]

The experience "crushed my spirits for a while," she says. After she was exonerated, she fled city politics and moved to San Cristóbal de las Casas, in the Mexican state of Chiapas, where for a few months she sang in the house band at a local disco. But performing six nights a week in a building she recalls as a hellacious firetrap couldn't match a safe desk job and a comfortable government pension. In 2003 she randomly ran into a city supervisor while hiking in the Guatemalan jungle and, soon after, was lured back to the Bay, first representing the SFMTA in the city attorney's office and then taking over the taxicab commission when it moved inside the MTA as part of the periodic reshuffling of the bureaucracy. She figured taxis would be fun and laid back. "Taxis were going to make elections look easy," she says. She quickly learned otherwise.

Hayashi got a close look at the city's taxi system, which had a fifteen-year waiting list for medallions, caps on the number of cars allowed, and nonexistent service outside of downtown and the airport. Everyone knew the taxi rules needed to be changed but no one could agree on how to do it. In 2009, Mayor Gavin Newsom asked her to overhaul the medallion system for the first time in thirty-two years and to install a New York City–style auction process to raise capital for the city. Hayashi worried that an auction would make medallions unaffordable for most drivers and came up with a new set of rules that raised the price to $250,000, offered low-interest loans to drivers, and gave older drivers a way to cut back on their hours.

Many other proposed changes during those years were bitterly contested by drivers and followed the exact same script as other taxi dramas playing out around the country and the world. Taxi drivers resisted any attempts to increase the number of medallions, reasoning that it would cut into their income and further congest airport taxi lots and the streets outside tourist hotels. They also vigorously fought efforts to mandate credit card readers in cabs because the transaction fees would come out of their pockets, plus their income would be documented and reportable to the government. Hayashi pointed out that they would certainly make up for this with increased tips, plus, passengers *wanted* to use credit cards instead of cash. They responded by encircling the MTA building and honking their horns in protest. One driver held a sign over his sunroof that read CHRISTIANE GO—LEAVE US ALONE.

Hayashi wielded a fast wit and plenty of personal charm to deal with cranky cab-industry vets who were hostile to change. But the fighting took its toll; she says she was "badly battered" by the credit card and medallion fracases and started to see her work as thankless. "I always joked my job was safe because no one else wanted it," she says. "The drivers hate you because their wife doesn't love them and their children are ugly and it's all your fault. The taxi-fleet managers don't like you because they aren't making any money. And any regulation is too much regulation."

She's talking about this years later, at a barbecue in the backyard of a friend's house in Berkeley. She's in her early fifties and visiting from Las Vegas, where she clerks for a county court and lives on a farm with a view of the mountains. Though she's recaptured her sense of humor, getting her to reminisce is difficult at first—those years at the MTA were the hardest of her life. "I was very stressed out in that job, which is one of the reasons why I love living in the country so much and not having any responsibility," she says.

In the summer of 2010, Hayashi's phone started ringing off the hook, and it wouldn't stop for four years. Taxi drivers were incensed; a new app called UberCab allowed rival limo drivers to act like taxis.

By law only taxis could pick up passengers who hailed them on the street, and cabs were required to use the fare-calculating meter that was tested and certified by the government. Limos and town cars, however, had to be "prearranged" by passengers, typically by a phone call to a driver or a central dispatch. Uber didn't just blur this distinction, it wiped it out entirely with electronic hails and by using the iPhone as a fare meter. Every time Hayashi picked up the phone, another driver or fleet owner was screaming, *This is illegal! Why are you allowing it? What are you doing about this?* She knew many of these drivers and fleet owners personally and had done her best to balance their interests along with the public's, but the result had been a system that didn't serve passengers or the city particularly well. Then Uber had radically tilted the entire playing field. The enraged drivers "were right," Hayashi says. "We are sitting here regulating the hell out of these poor guys and then we just ignore what was going on?"

Hayashi was aware of the limits of her authority. Regulating limos and black cars was the responsibility of the state, not the city. But she saw an opening: this startup was calling itself Uber*Cab* and thus seemed to be marketing itself as a taxi company. She talked to the enforcement division at the Public Utilities Commission of

California, which was tasked with regulating limos and town cars, and they orchestrated the joint cease-and-desist. After receiving the threatening letter, Uber promptly asked for a meeting.

Travis Kalanick, Ryan Graves, and Uber's outside lawyer Dan Rockey met Hayashi and other city and state officials on November 1 in a conference room on the seventh floor of 1 South Van Ness. It was the first of countless times that executives at Uber would face government officials to discuss the legality of their company's service. Graves says they were nervous. "We didn't know what to expect," he says. Beforehand, the Uber team agreed on a respectful, inquisitive, cooperative, and confident tone.

Somehow, things fell apart anyway. Kalanick later said the PUC officials were reserved and asked for more information but that Hayashi "was fire and brimstone, deep anger, screaming."[3]

Hayashi says that she was strident, not screaming, and remembers the Uber execs as "obnoxious" and Kalanick in particular as "arrogant." "You can't do this!" she told them. "You can't just open a restaurant and say you are going to ignore the health department!"

She says that nothing was decided at the meeting and calls it "totally pointless." But that's not entirely true. In fact, Uber's first clash with city regulators likely changed the course of this tale.

Garrett Camp had been trying to get his friend Travis Kalanick more involved with Uber for almost two years. From the mad sprint on the morning of Barack Obama's inauguration to their adventures at South by Southwest in Austin, the Lobby conference in Hawaii, and LeWeb in Paris, Camp had been evangelizing for a world in which luxury cars could be summoned with a tap on a smartphone. That fall Kalanick was working at Uber a few days a week, signing up limo fleets, and leading many conversations with investors, and he did much of the talking in the critical meeting with Hayashi and the other regulators. Uber remained a side project. Ryan Graves was still the CEO. But slowly, steadily, Kalanick had started to believe.

Kalanick was still in his self-described "burnout phase" after his last full-time job.[4] He was traveling around various countries in Europe and South America, wearing a dorky cowboy hat; when back at home, he applied his capacity for manic focus to mastering video games like Wii Tennis and Angry Birds. Chronically restless, he was also investing and advising various startups and giving occasional speeches about his past misadventures as an entrepreneur.

Camp knew that Kalanick would be perfect for Uber. His friend loved digging into the details of complex businesses and plumbing the secret science of building startups. So Camp, still consumed with his newly independent first company, StumbleUpon, continued to press Kalanick to take over Uber. "I really think Travis should run it," Camp said to one of Uber's earliest advisers, Steve Jang, that year. "He's almost there. He's close."

By the time of the first fateful meeting with Hayashi, Kalanick was telling friends he was ready to find a new full-time job. But it wasn't necessarily going to be at Uber. Another company he was advising, Formspring, a question-and-answer site that had raised $14 million, seemed poised to become the next big social website and was negotiating with him to be its chief operating officer. Ade Olonoh, the co-founder of Formspring, says that discussions got so advanced that Kalanick was offered the role and there were board conversations about Kalanick's compensation. Kalanick told me it was one of several jobs he considered at the time.

Formspring was one of ten companies in which Kalanick had made angel investments. He fashioned himself a hands-on mentor to young CEOs, the Silicon Valley version of the Wolf from *Pulp Fiction,* who could drop into tricky situations and help raise money or negotiate deals.[5] "His skill was taking a messy hard problem and being a facilitator, willing and ready to roll up his sleeves," says Olonoh. "He had a lot of pride in being the type of investor who helped his companies."

Kalanick discovered another startup, CrowdFlower, which relegated menial business tasks to independent workers over the internet, by cold-calling its customer-support line and striking up a friendship with its CEO, Lukas Biewald. For two years, they spoke a few times a week, and Biewald was a frequent guest at the Jam Pad. "He helped me when he had no reason to," Biewald says. Kalanick was full of tips on how to deal with investors, hire top execs, and negotiate with prospective partners. "Lukas, everyone is going to give you advice," Kalanick told him. "Ask for the story behind the advice. The story is always more interesting."

Travis Kalanick was born in 1976 and raised in a home on a leafy street in the middle-class suburb of Northridge in L.A.'s San Fernando Valley. His father, Don, served two years in the U.S. Army and was a civil engineer for the City of Los Angeles. His mother, Bonnie, sold ads for the *Los Angeles Daily News.*

At Granada Hills High School, Kalanick ran track, anchoring the 4-by-400-meter relay and specializing in the long jump.[6] A photo in his high-school yearbook shows him in midjump, his right leg outstretched, his face clenched with focus. "I would put it all in," he said. "Leave it all on the field."[7] He spent one summer driving his old '86 Nissan Sentra around to neighborhood homes and selling twenty thousand dollars' worth of knives for Cutco, a brand of kitchen merchandise often hawked door to door by students. Occasionally he was the butt of his friends' jokes; they often commented on his "sharp" attire.[8]

Talented with numbers, Kalanick got a perfect score on the math portion of the SAT and became a neighborhood tutor. "I could get a thirty-minute math section done in eight minutes," he said. "You put me on the verbal [SAT test] and my shoulder would hurt, my neck would hurt, I'd take the entire thirty minutes and I'd be stressing. But math I'd just fill in the bubbles."[9]

The summer after he graduated high school, he started an SAT training company called New Way Academy with a classmate's father, a man who belonged to the local Korean church. They advertised the courses through the church, and hundreds of kids signed up. Each Saturday morning during his freshman year at UCLA, Kalanick would put on a white shirt and a tie and teach a class called 1500 and Over. The name itself was a sales pitch to students and their parents. "The first person I tutored went up by 400 points," he bragged years later.[10]

Kalanick lived at home during college and pursued a computer science degree. But this was the late 1990s, and for those whose interests lay at the intersection of entrepreneurship and computers, the siren call of the internet was irresistible. Kalanick dropped out of school his senior year, 1998, to join six classmates developing one of the web's first search engines, Scour.net. The site, which debuted around the same time as Google, let people search the computers of other students on university networks for multimedia files like movies, TV shows, and songs. Most of those files, of course, were being hosted and downloaded online for free, in violation of copyrights.

In the site's first year, the *L.A. Times,* the *Wall Street Journal,* and numerous other publications wrote about the company, and user growth took off. The seven classmates holed up a two-bedroom apartment near fraternity row and worked, ate, and slept there. "Anyone with any concept of hygiene would just be offended by what was going on in that place," says a co-founder, Jason Droege, who would later join Kalanick at Uber.

Scour was a hit on college campuses; in June 1999, the Scour website was getting 1.5 million page views a day and had logged 900,000 visitors over the previous two months.[11] Kalanick, the oldest member of the group and a self-styled businessman among coders, was vice president of strategy, in charge of cultivating investors and media partners. Droege recalls that a twenty-two-year-old Kalanick already had a penchant for constant pacing, his phone usually pressed to his ear, focused totally on finding anyone who could possibly help the young startup.

Discussing these early startup experiences, Kalanick would later refer to himself as a chronically "nonlucky" entrepreneur—someone who labored for years but never seemed to get any breaks. That history of hardship started here, in the Wild West days of online deal-making. In 1999, Scour was set to raise millions from superagent and former Disney president Michael Ovitz and the investment firm of supermarket mogul Ron Burkle. The notoriously aggressive Ovitz, who wanted to augment his e-commerce website Checkout.com with a network of other internet properties, set about trying to maximize his leverage, stringing out negotiations for nine months after the parties had agreed in principle to a deal. When the impatient founders finally tried to solicit other investors, Ovitz sued Scour in the Los Angeles Superior Court, alleging that they had reneged on the agreement.[12]

When the smoke cleared, Ovitz and Burkle had acquired 51 percent of the company,[13] and the young, impressionable Scour founders had gotten a valuable lesson in the brutality of business in the big leagues. Nevertheless, Scour flourished, at first. The founders moved into Ovitz's swank offices in Beverly Hills, eventually hired seventy employees, and got an education in the L.A. business scene, reading books the Ovitz crew passed them, like *The Art of War,* by Sun Tzu, and *The Forty-Eight Laws of Power,* by Robert Greene. Kalanick and his colleagues believed they could work with rights holders to create a more efficient and economical way to distribute media over the internet. When the rogue file-sharing service Napster took Scour's technology a step further, allowing people to not only search for files but pass them back and forth, Scour moved quickly to catch up. It introduced its own version of the technology, called Scour Exchange, which made it even easier to trade audio and video files without paying.

Then Hollywood woke up to the impact of peer-to-peer file sharing and moved swiftly to crush it. Kalanick and his colleagues had met with all the leading music and movie studios and believed that the meetings had gone well. But in July of 2000, thirty-three media companies, including the major music and movie trade groups, sued Scour in court for a whopping $250 billion. "This lawsuit is about stealing," said legendary MPAA president Jack Valenti. "Technology may make stealing easier, but it doesn't make it right."[14]

Scour's allies ran for cover. Even Ovitz backed far away and, Kalanick later asserted, had a colleague threaten Kalanick with physical harm if he dragged Ovitz's name any farther into the fracas.[15] Ovitz denies he ever threatened Kalanick and talks about him favorably as a young but impressive negotiator who couldn't quite see the larger picture when the industry rallied against file sharing. "Travis didn't understand that we had made a mistake" in backing Scour, Ovitz told me in 2015 at a tech-industry conference. "We didn't realize we were creating enemies in the world of intellectual property. If you got sued by every angry music and movie company and everyone in the world who has IP, you'd notice. That didn't bother Travis. It sure as hell bothered me."

Scour's attorneys, like Napster's, believed the company was protected by the "safe harbor" provisions of the Digital Millennium Copyright Act of 1998, which stipulated that internet companies could not be held liable for the activities of their users. Scour, they argued, wasn't hosting the content, only pointing to it. But the startup couldn't hope to fight the combined might of the entire media industry. It laid off most of its staff in the fall of 2000 and declared bankruptcy to escape the litigation.[16] "That's when we really learned how the world can work," says Droege. "It's not whether or not you are right or wrong."

In bankruptcy court, after a fifteen-minute auction, the assets of the company were sold for nine million dollars to a little-known Oregon firm.[17] Kalanick, still only twenty-four, had to watch as everything he had worked for, the dream he had quit college to pursue, was trampled by powerful companies and their high-priced lawyers.

It was the kind of traumatic experience that could harden the character of a young entrepreneur. It was also just plain depressing. "By the time we actually truly went out of business, I was probably sleeping fourteen hours a night," he said later.[18] In public, he tried to hold his head high. "I was [playing] the game I call 'fake it till you make it.' Basically fighting reality. When you do that too long, when you are in failure state, it will eventually crush you."

Despite that setback, Kalanick was ready to dust himself off and try again.[19] He started talking to one of his Scour co-founders, Michael Todd, about redeveloping the technology behind Scour and selling it to media companies as a tool to help them distribute their material online. Bandwidth was expensive back then, around six hundred dollars per megabyte (as opposed to about a dollar per megabyte on a broadband internet line today), and peer-to-peer networking could reduce the cost. They called their new company Red Swoosh, after the twin half-moon insignias in the original Scour logo. Kalanick said it was "a revenge business" and recognized a satisfying irony: "The idea is the same peer-to-peer technology but I take those thirty-three litigants that sued me and turn them into customers," he said. "Now those dudes who sued me are now paying me. It sounded good."[20]

In practice, it didn't work out as well.

Kalanick tried to raise money in 2001, right in the midst of the dot-com bust. Silicon Valley was a ghost town. At a local bar in Palo Alto, a venture capitalist told him that all innovation in software had been done and there was nothing left to invent.[21] On September 11, he had a meeting scheduled in L.A. with Daniel Lewin, a co-founder of the Boston-based streaming-media company Akamai. Lewin was on American Airlines Flight 11 and died in the terrorist attacks.

Red Swoosh had an office in Westwood, seven full- and part-time employees, mostly refugees from Scour, and a few paying customers. But the scent of failure was strong, even at the beginning. Todd and Kalanick couldn't agree on company strategy, and bandwidth prices were dropping and making the product less compelling. Kalanick claimed he discovered that Todd wasn't properly withholding the firm's payroll taxes and was trying to surreptitiously sell the engineering team, without Kalanick, to another company.[22] Todd left Red Swoosh over the spat and disagrees with Kalanick's version of events, saying only that "Travis is a great storyteller."

Todd landed a job at Google and promptly hired away Kalanick's last software engineer. Kalanick, twenty-seven, was now utterly alone. He had lived at his parents' house for a year and had gone without regular paychecks while pursuing deals with companies like Microsoft and AOL, only to watch them invariably fall through. "Imagine hearing 'no' a hundred times a day for six years straight," he told me years later. "At some point even your friends are like, 'Dude, you need to do something else.' To keep going in the face of that can be a lonely existence."

Kalanick tried some unusual tricks to get attention. In 2003, while at the county recorder's office in Hawthorne to get a passport, he noticed television news trucks parked out front. Curious, he asked why they were there and learned they were covering prospective candidates who were registering to run in that year's election to recall and replace California governor Gray Davis. A self-described C-Span junkie in high school, Kalanick was intrigued and put himself on the list to run. He then spent a few days canvassing Hermosa Beach near his home, telling sunbathers about his file-sharing platform and trying to obtain the needed ten thousand signatures to get himself on the ballot. He got about fifteen. "I only had certain things to say, you know. I didn't have much," Kalanick recalls.

Kalanick may not have considered himself a serious candidate for governor but he stubbornly believed he could make Red Swoosh work. Internet mogul Mark Cuban saw promise in the idea and, despite the fact that Kalanick had no regular employees, invested a million dollars in the business in 2005. It was enough to keep going. "I like to call these my blood, sweat, and ramen years," Kalanick said. "I always very much believed in what we were doing."[23]

With the new capital, there was only one thing to do: move to Silicon Valley. He found a small office in San Mateo, twenty miles south of San Francisco, and, utilizing only his conviction and charisma, he hired four engineers. David Barrett, the first to sign on and the future founder of a cloud software company called Expensify, says Kalanick was "completely honest about the state of business" and was "persuasive, compelling, and candid." He found Kalanick's enthusiasm infectious. "If you had a shit-ton of data, we gave you a way to move it," Barrett says. "The problem was that there were only three companies in the entire world who wanted to do it."

Enjoying a modicum of momentum, Kalanick leased a new office in San Francisco but had a month before he could move in. Instead of waiting, he took the whole company to Thailand, where they worked eighteen-hour days out of cafés and a house overlooking the craggy Railay Beach coastline rewriting the Red Swoosh code. It was a productive retreat and the first of what Kalanick called *workations,* a tradition that continued at Red Swoosh and, later, Uber.

Back in the Bay Area, Kalanick raised more money from August Capital—the firm that later passed on Airbnb—and resumed scrapping to find a graceful exit for Red Swoosh. Again demonstrating his skill as a salesman, he landed satellite TV provider EchoStar as a client and, in 2007, sold the entire company to Akamai for $18.7 million, plus an extra payout if the company met certain goals.[24] It was a meager exit by most Silicon Valley standards but a huge relief for Kalanick, netting him several million after six years of deprivation and anonymous toil. "He could have and should have given up well before he sold it," says David Hornik of August Capital. "He deserved it."

Kalanick had endured the most grueling experience of his life and emerged as battle-hardened and defiant as ever. Around this time, he went out to a nightclub in San Francisco with several friends, including Napster co-founder and Facebook investor Sean Parker. At the end of the night, inebriated, Kalanick was waiting for his friends outside the nightclub when an imperious bouncer told him to move away from the door. Kalanick moved only a few steps away. "Keep going," the bouncer ordered. Kalanick inched over another step. "Keep going," the bouncer said menacingly. "I'm not breaking the law. You tell me how I'm breaking the law," Kalanick replied.

By the time a nearby police officer arrived, the bouncer was forcibly trying to move Kalanick while he defiantly gripped a parking meter with both hands. He was arrested for obstruction of the sidewalk and says he spent eight to ten hours in the city jail before Parker realized what had happened and put up two thousand dollars to bail him out.[25]

"Fear is the disease. Hustle is the antidote," he said at a Chicago startup event a few years later.[26] "You start a company in 2001, good luck, right? You can't count on funding. You can't count on sales. You can't count on anything but just crazy hustle and just grit your teeth, claw your way to success. There was just no easy way to do it."

To commemorate the sale of Red Swoosh, Kalanick bought a pair of patterned socks with his new motto embossed on them: "Blood, sweat, and ramen."

Now he had a decision to make. His friend Garrett Camp wanted him to take the top spot at Uber. But in 2010, Uber was a tiny company. It had some half a dozen employees, a few dozen limo drivers in San Francisco using the platform, and little in the way of expansion plans. Its motto, "Everyone's private driver," conveyed luxury and exclusivity, not mass-market appeal. And Kalanick was reluctant to displace Ryan Graves—or at least savvy enough to know that Silicon Valley in the age of venerated founders like Mark Zuckerberg tended to look askance when investors ousted the original chief executive.

At the same time, Uber excited him in a way that Formspring, the much larger Q-and-A site that was also recruiting him, did not. Uber was turning out to be a company rooted in complex math. Its biggest challenge, and where he found himself already frustrated with the performance of the startup, was finding ways to attract more drivers during peak times and to route cars into the areas of highest demand. Uber had the data to make those kinds of prescient decisions. In fact, it was slowly dawning on the founders and board members that Uber was going to have more data about how people moved around cities than just about any other company in history. "I'm an engineer at heart and math moves the needle on this," Kalanick told me a few years later. "My happy place is right in the midst of all that complexity."

Uber's financial results also looked promising. The company was exhibiting an elusive phenomenon called negative churn, in which users who joined the service were more likely to stay with it and gradually increase their frequency of use than they were to leave. In other words, once customers joined Uber, they turned into a sort of high-yield savings account. The lifetime value of a user seemed unknowable, perhaps unlimited. When someone signed up, according to an early internal estimate, he or she was worth forty or fifty dollars a month in gross revenues and eight to ten dollars in gross profit—for the foreseeable future. "This is the equivalent of a perpetual-motion machine and cannot go on forever, but it means that rider spending is accelerating at a rate that is bigger than our churn rate," Kalanick wrote to his fellow investors in an e-mail that year.

These kinds of numbers were rare in a startup. They were the sort of numbers that could attract significant new financing and fuel rapid expansion. Uber might be the canvas on which Kalanick could apply all of his talents, hard-won experience, and ambition.

But that alone didn't tip the scales. In speeches that fall, Kalanick was still talking about himself as a "startup consigliere" and "the Wolf."[27] Then he had that explosive meeting with Christiane Hayashi.

The meeting thrust Kalanick back into the thick of the familiar battle between new technology and the old, outdated ways of doing things. In the weeks after, he kept the Uber board informed about negotiations with the city. Uber had to stop marketing itself as a cab company, but that was an easy compromise. By the time of the cease-and-desist, the founders had already decided to drop the *Cab* from the UberCab name, and investor Chris Sacca was negotiating with the Universal Music Group, which owned the Uber.com web address, to buy it for a 2 percent ownership stake of the company (at the time, about a hundred thousand dollars). Universal Music also elicited a promise from the startup that if Uber didn't work out, the company would give the name back.

The California Public Utilities Commission wanted Uber to register itself as a limo company or, technically, as a "charter party carrier," but Uber's lawyers believed the company could make the case that it was merely an intermediary between drivers and riders, not an actual fleet operator. It was as much a limo company as Orbitz or Expedia were airlines, they argued. In a follow-up ruling in late 2010, the PUC agreed, and Uber never stopped operating. To the consternation of Christiane Hayashi, who was trying and failing to get the city attorney's office to give her authority to regulate the startup, Uber had a winning argument.

Kalanick later said that Uber's first fight in San Francisco added to his personal conviction about the company just as he was taking a more active leadership role. "For me that was the moment where I was like, for whatever reason, I knew this was the right battle to fight," he told me in 2012. On a tech podcast, he added that the fight with the MTA recalled all the litigation and conflict of his decade in the world of peer-to-peer technology. "The great thing is I've seen this before," he said. "I thought, *Oh, man, I have a playbook for this. Let's do this thing.* When that happened, it felt like a homecoming."[28]

After that first meeting with Hayashi, Kalanick spent weeks negotiating with Garrett Camp and angel investors Chris Sacca and Rob

Hayes over his compensation as CEO. He somehow calculated that he needed a 23 percent ownership stake in Uber, up from his 12 percent stake as a founder and adviser, but declined to explain his logic. The other board members didn't want to dilute their own holdings but ultimately acquiesced. "The best thing I ever did for Uber was completely roll over in negotiations with Travis," says Rob Hayes.

Finally, Kalanick himself delivered the news to Ryan Graves, pitching it as a partnership and the opportunity to work together more closely. If he was angry or upset at the demotion, Uber's original CEO hid it well. "I kind of revisited what I was hoping to get out of the experience in the first place," says Graves, whose title changed to general manager and later, vice president, operations. He recalls telling Kalanick, "As long as it's a partnership, as long as it doesn't feel like a job, I'm good. I didn't come here to take a job. I had gotten fully sold on the idea of running a company. As long as that's the case, then all good. I trust you."

They signed the final documents on November 23, 2010, and announced it to tech blogs a month later.[29] Graves wrote on Uber's website that he was "superpumped" about having Kalanick on board full-time, deploying a gung ho phrase that would become a motivational motto among future Uber employees. And Kalanick expressed some of the hard-charging enthusiasm and ambition that he had brought to Scour and Red Swoosh.

"The bottom line is that I'm all in on Uber," he wrote.

> The excitement and joy of being Uber is coming out my pores and I'll stop at nothing to see Uber go to every major city in the U.S. and the world. So what's next? Taxi frustration is going down. Reliability, Efficiency, Accountability, and Professionalism in urban transportation are going way up. Every city Uber rolls into is going to be a better place when we're done with it and if you live in that city, your world of transportation is changing forever, and it will be oh so Uber when that change arrives.

* * *

There was another unintended consequence of Christiane Hayashi's fight with Uber. The young startup received a torrent of publicity from Silicon Valley's technology blogs. Between that and the already strong word of mouth, the number of Uber rides started bounding upward by 30 percent each month.

The company had moved temporarily into the local offices of First Round Capital, where they availed themselves of the foosball table and other perks of venture capital life. Every few days, Kalanick would charge excitedly into Rob Hayes's office with new data. Once he ran in and announced that thirty-five rides had been completed in a single hour! A new record! "I remember looking at Travis and saying, 'Dude, I think you have the tiger by the tail. I think this is real,'" Hayes says. "He just smiled devilishly at me."

Kalanick was now prepared to devote himself fully to another entrepreneurial adventure. He stopped angel investing, curtailed his advising of other companies, and even broke up with a longtime girlfriend, explaining to a stunned colleague, "I realized I was more passionate about this company than I was about her. I should probably find someone I like at least as much as my job." He also showed early flashes of belligerence toward competitors—an augur of the conflicts to come. "They will be getting into one of the most complex businesses I've ever seen for all the wrong reasons, and they will sorely underestimate the pummeling they will go through at the expense of my bare knuckles," he e-mailed a friend who was pointing out a Tweet critical of Uber by a potential rival. Kalanick signed off with *Bleeding Uber Blood.*

First, though, Kalanick had to address some of Uber's early problems. At the time, the Uber app was showing users how many cars were available in their area. Users opening the app and seeing no cars available, an all-too-frequent occurrence the founders started calling "zeros," had always exasperated Kalanick and Camp. That, in their minds, was not an "Uber experience." To fix it, Uber had to add new drivers, predict when and where spikes in demand would occur, and encourage drivers to flock to those neighborhoods.

This required a wholesale identity change. Uber, Kalanick realized, wasn't really a lifestyle company that offered swank rides, despite what the motto "Everyone's private driver" suggested. It was a technology company, and it had to become intimately familiar with its own internal metrics. "This is a company that needs to run on data," Kalanick told colleagues. He amped up hiring, and in December he recruited a new director of engineering, Curtis Chambers, whom he had first met at Akamai. Chambers started work on a new dispatch system to replace the already creaky original created by Oscar Salazar, which was being held together with spit and duct tape by engineer Conrad Whelan.

The Uber execs were dealing with seasonal swings in demand, which they were seeing for the very first time. The daily, weekly, and seasonal rhythms of the city's transportation grid were beginning to present themselves. Halloween had been busy, while Thanksgiving, to their alarm, was slow—people stayed at home, it turned out. Over the Christmas holiday, anticipating a rush on New Year's Eve, they made their first attempt to bring supply and demand into balance. Uber enlisted as many drivers as it could and raised the fares to twice the usual rate for that night. It also ran a lottery system and awarded a handful of users VIP status, which locked them into the normal fare and gave them exclusive access to a few dozen cars. Then Kalanick and some of the engineers decamped to Marina del Rey in Los Angeles to watch it all unfold. It was Uber's first workation.

Amid overloaded servers, intermittent service, and a balky app that badly needed an upgrade, the first New Year's experiment didn't work that well. But it started the company down a path—a controversial one, it turned out—of using price as a way to deal with volatile demand.

As 2011 began, Kalanick had another big move in mind. It was time for Uber to raise its first significant round of funding, the Series A. He wanted to work with one investor in particular: Benchmark's Bill Gurley, who had previously expressed interest in the seed round.

Gurley had tracked Uber's progress closely over the nine months since then, what the former Florida Gators basketball player calls "hanging around the rim." Sensing the opportunity to bring transportation online in the same way OpenTable had consolidated restaurants and Zillow had aggregated real estate listings, Gurley was aggressive. He went on a bike ride with Chris Sacca in Truckee to talk about the company and drove up to San Francisco late one night to spend two hours with Kalanick at the W Hotel bar, hammering out prospective deal terms.

Gurley had identified a big opportunity but he was also fortunate. He had tried and failed with Taxi Magic and Cabulous, two investments in rival companies that would have precluded his backing Uber. Now he recognized that Uber, free from the regulation and price controls that governed the operation of yellow cabs, was the larger prize.

Benchmark almost scuttled the deal with a practical joke. Kalanick was on Sand Hill Road in Menlo Park to visit rival Sequoia Capital before a scheduled meeting with Benchmark's partnership. As they waited for Kalanick to arrive, Gurley and his partner Matt Cohler looked at the Uber app and saw a single Uber car in front of Sequoia's office a mile away. Because Uber did not yet operate down in Silicon Valley, they guessed this oddly idle car was Kalanick's ride. Cohler summoned the car with the Uber app on his phone, and when Kalanick came out of his meeting, it was gone. He had to run over to Benchmark in his dress shoes, and he arrived sweating and late. That night, the firm sent Kalanick a pair of running shoes. "I don't know why we thought that was a good idea," Gurley says of the prank.

Kalanick didn't hold a grudge, and Benchmark led an $11 million fund-raising round that valued the fledgling startup at $60 million. Rivals like Sequoia and Battery Ventures, which had considered investing in the round and passed, added their names to the list of everyone who underestimated either the size of the opportunity or the fortitude of its new CEO. "Scour and Red Swoosh were tough," Gurley says. "All of the sudden, Travis had a little wind at his back. It often felt like he thought he had an obligation to the entrepreneurs' society of the world to play Uber for all that it was worth."

Kalanick, the combative CEO who had something to prove in the wake of his past business failures, and Gurley, the seasoned investor who intimately understood the benefits and challenges of building an internet marketplace, were a potent combination. With fresh capital in the bank, they agreed, Uber needed to do one thing right away—expand out of San Francisco and into every major city in the world.

PART II

EMPIRE BUILDING

CHAPTER 6

THE WARTIME CEO

Airbnb Fights on Two Fronts

The Peacetime CEO does not resemble the Wartime CEO.

—Ben Horowitz, *Ben's Blog*[1]

The pall hanging over Silicon Valley from the financial crisis started to lift in 2011. Facebook led the way in January, raising half a billion dollars from an investor group headed by Goldman Sachs after announcing that it had signed up more than five hundred million users. The professional networking site LinkedIn went public in May, scoring a four-billion-dollar market capitalization. Though the term wouldn't be coined for another few years, this was now the age of the unicorn—not Sofiane Ouali's white 2003 Lincoln Town Car but tech startups valued at more than a billion dollars.[2] That year the streaming-music service Spotify, the cloud storage company Dropbox, and the payments startup Square all became charter members of a club that everyone would soon want to join.[3]

The smell of optimism was in the air again, along with the belief that a well-timed internet startup could ride a great wave of converging technology trends.

This change in sentiment was lubricated with capital. The bonds markets were flat and the stock market was lifeless, but venture capitalists, touting returns from the previous generation of upstarts, could still beguile investors with fantastic dreams of rapid growth and wealth creation. The Russian venture capitalist Yuri Milner of Digital Sky Technologies, or DST, had been ridiculed a few years before for investing $200 million in Facebook for a 2 percent stake. That March he bought a house in Los Altos Hills, an eighteenth-century French château–style mansion with panoramic views of the golden San Francisco Bay. In Silicon Valley, this is what passes for getting the last laugh.

To the casual observer, the home-rental site Airbnb didn't seem like it could ride this wave, let alone come to embody it. At the start of the year, its employees were still crowded into the office on Tenth Street in SoMa (the South of Market area in San Francisco), with bad cell phone reception inside and the homeless camping on the street outside. The startup was run almost entirely by its triumvirate of co-founders, who had two college degrees in design and one in computer science among them.

Behind the scenes, though, Airbnb was booming. The activities of Nate Blecharczyk, the growth hacker, had cranked the flywheel. Ample coverage in the media, thanks to Brian Chesky and Joe Gebbia's charismatic recounting of the company's history, turned it ever faster. Hosts brought a variety of eclectic properties onto the site, from two-bedroom homes in newly hip Venice Beach to castles in southern France, tree houses in Northern California, and the fuselage of a retired 727 Boeing jet in Costa Rica.[4] Because of the convenient fact that guests were *travelers*, word of mouth was rapid and global—as viral as a potent strain of the winter flu.

For Chesky, the CEO, Airbnb's emergence was a ticket into the rarefied business elite. In March he was invited to speak at an Arizona technology conference held by the investment bank Allen and Company, where he enchanted the audience with the narrative of Airbnb's unlikely origin, from the design conference to the cereal gambit. A few months later he was invited to the investment bank's annual gathering of the rich and famous in Sun Valley, Idaho. Now he was rubbing shoulders with Oprah Winfrey, Warren Buffett, and Bill Gates. At one point he found himself explaining the home-sharing idea to the actress Candice Bergen; he kept thinking, "Murphy Brown came up to me. Murphy Brown knows Airbnb," Chesky says. "It was this crescendo. It was an airplane, reaching a higher and higher elevation."

In May, Chesky met fellow traveler Travis Kalanick for the first time. At a conference in New York City called TechCrunch Disrupt, Chesky and Kalanick were invited to appear onstage together in a panel discussion titled "Disrupting Offline Businesses." Chesky had been an Uber fan since Ryan Graves had invited him to have coffee in 2010 to solicit advice about running a startup, and he had turned his employees into avid Uber users. Kalanick had once thought about starting his own home-sharing business, Pad Pass. So the two had plenty to talk about. The night before the conference, Kalanick e-mailed Chesky out of the blue, suggesting that they "get together and jam." They met for dinner in midtown Manhattan. Chesky found the Uber CEO relaxed and personable.

The next day, though, a far more provocative and cocksure Kalanick showed up for their joint interview wearing pink socks. Moderator Erick Schonfeld from *TechCrunch* asked Chesky about reports that Airbnb was raising a massive round of funding that could value it at a billion dollars—unicorn territory. "Can't comment on that, unfortunately," Chesky said.

"Why would you deny a billion-dollar valuation?" interjected Kalanick, drumming his thumb against his leg. (Uber's valuation at the time: only $60 million.) "Just roll with it."

Chesky cast him what seemed to be an incredulous sideways glance.

Schonfeld noted that both CEOs had survived initial tangles with local governments. The cease-and-desist against Uber in San Francisco had been dropped, though Kalanick still exaggerated it for theatrical effect. "I think I've got like twenty thousand years of jail time ahead of me," he said, to significant applause. Chesky, perhaps considering the audience beyond Disrupt, insisted that "the spirit of the local governments essentially support Airbnb,"[5] and he downplayed a recently passed New York state law that forbade people in New York City to rent out their homes for less than thirty days.

The two had plenty in common. Kalanick, then thirty-four, and Chesky, twenty-nine, were young CEOs at the vanguard of renewed Silicon Valley optimism; they were confident, charismatic, and unaware of the looming conflicts with rivals and regulators. As the world opened up to them, each would pursue the opportunity aggressively, occasionally ruthlessly, and with varying degrees of moral rectitude. This was a time to build empires.

Though he wouldn't acknowledge it, Chesky was in fact wrapping up a monster financing round. He had started looking for new capital that spring and found a receptive audience. With Airbnb bookings growing 40 to 50 percent every month, *TechCrunch* had called it "the sleeper hit of the startup world."[6] Andreessen Horowitz, which passed on the Series A, beat out a crop of other top-tier venture firms in a hotly competitive fund-raising round. It led the Series B and a group that included Yuri Milner's DST, the personal investment funds of Amazon founder Jeff Bezos, and the actor Ashton Kutcher for a total $112 million investment that valued the company at $1.3 billion.

The financing was led by Andreessen Horowitz's Jeff Jordan, a former eBay president. Jordan had gone from thinking "this is the stupidest idea I've ever heard" to practically jumping out of his seat at the Allen and Company conference when he recognized the similarities between Airbnb and his old company.

"The community had taken a small idea and made it into a huge idea, just like at eBay," says Jordan.

Despite this optimism, Jordan and his partners identified four risks to their investment:

Safety: What would happen if a guest trashed a home or apartment?

International competition: Would overseas entrepreneurs clone the site?

Regulation: Would cities allow hosts to continue to rent their homes without restrictions?

Executive recruitment: Chesky, Gebbia, and Blecharczyk were running the company as a triumvirate—a council of equals. It was an arrangement that couldn't last. Could they find new executives they trusted?

All these critical concerns would soon prove valid. "Nothing we do is riskless. That's why it's called risk capital," Jordan says. "This thing obviously had a really interesting upside but it also had some hair on it. Brian knew what the hair was."

The most pressing question of all was whether he would be ready.

That spring, Airbnb engineers noticed unusual activity on its website and mobile apps. Automated software programs were visiting the site and collecting or "scraping" the personal data posted by hosts. Soon after, the company started hearing that hosts in Europe were being contacted by phone, e-mail, and even in person by salespeople for other home-sharing websites. It didn't take long for Airbnb to realize what was happening. The clones were coming.

Most successful internet startups are copied by opportunistic entrepreneurs around the world. Airbnb was no exception. The first clone, called 9flats, emerged in February of 2011. It was based in Hamburg and had been founded by Stephan Uhrenbacher, who had also founded a company called Qype, a copy of the American reviews site Yelp. Uhrenbacher raised around $10 million for 9Flats (motto: "Stop being a tourist. Feel at home in the world") and said that he too wanted to become a "global player" in the online travel industry.[7]

Another copycat company was launched in April and had a far

larger impact back at Airbnb headquarters. Wimdu, based in Berlin, was founded, funded, and operated by the Samwers, the fearsome trio of German brothers that Brian Chesky had been warned about. Wimdu resembled Airbnb right down to the light blue color scheme and the search bar, which asked "Where do you want to go?," only a slight variation of Airbnb's "Where are you going?" In a brazen flourish, Wimdu advertised on the bottom of its home page that "the concept" behind the site had been featured on CNN and by the *New York Times*. Those outlets had written about Airbnb, not Wimdu.

Marc, Oliver, and Alexander Samwer, then all in their late thirties, had grown up in Cologne, Germany, sons of two corporate lawyers who had their own private practice. The brothers were close from an early age and searched for ways to combine and exploit their joint talents. After getting degrees in law (Marc) and management (Oliver and Alexander), they moved to Silicon Valley and took jobs among the first generation of internet companies, not to embark on careers in the U.S. technology industry but to watch and learn. After returning to Germany in 1999, they started a German-language auction site called Alando that looked and operated like eBay. Alando established a foothold in Germany and after four months, eBay bought it for $43 million, making the Samwers millionaires. It was a start.

Over the next decade, the Samwers founded and invested in companies that emulated Facebook, eHarmony, Twitter, Yelp, Zappos, and YouTube, then made billions by selling each one off, often to the company they'd copied. They didn't apologize for this, noting that "BMW didn't invent the car" and that it was the execution, how you built and operated a startup, that really mattered.[8] They worked grueling hours, crisscrossing the globe, hiring and making deals with blazing speed. Stories about their peculiar work habits were legendary. When Marc Samwer took long flights, colleagues said, for exercise he would fully recline his airline seat and bicycle-kick in the air for thirty minutes. Oliver, they said, traveled the world with only a small briefcase that held exactly one pair of underwear and a clean shirt. Every morning he would wash the clothes he wasn't wearing in the hotel bathroom and leave them there to dry.

The Samwers brought military bravado and combat terminology to the art of creating internet startups. In a 2011 e-mail to colleagues at Rocket Internet, the brothers' startup incubator, Oliver Samwer wrote in his customarily stilted English to colleagues who were developing a furniture-selling website: "The time for the blitzkrieg must be chosen wisely, so each country tells me with blood when it is time... Now it is time to either decide we will die to win or to give up... i do not accept surprises. I want this planned confirmed by all three of you: you must sign it with your blood."[9]

The e-mail was leaked to *TechCrunch*. Samwer apologized for the tone and for using an infamous term from German military history.

Most U.S. startups approached competition from the Samwers with fear and a sense of futility, concluding that it was easier to roll over than fight. A year before they founded Wimdu, the Samwers started a Groupon clone called CityDeal, backing it with 20 million Euros from Rocket Internet and quickly turning it into Europe's leading daily-deal site. They competed fiercely with another European clone called DailyDeal. At one point the Samwers sent job offers to many DailyDeal employees, offering promotions and raises if they defected, according to a profile of the Samwers by Caroline Winter in *Bloomberg Businessweek*.[10] They also spread rumors that DailyDeal was close to bankruptcy. Oliver Samwer didn't exactly apologize for these tactics. "I think it's all within the normal laws of competition," he told Winter.

In 2010, Groupon acquired CityDeal for around $126 million and kept the brothers on to run it. It was an enormous mistake. After Groupon's 2011 IPO, the European division run by the Samwers suffered from chronic technology problems and alienated customers with two daily-deal e-mails instead of one. Oliver

Samwer and Andrew Mason, Groupon's CEO, constantly argued over whether sending multiple messages a day was a good idea, according to a former Groupon executive. Two daily deals generated more revenue but diminished the novelty and quality of the deals. Mason would be fired as the CEO of Groupon in 2013, partly due to ongoing problems with its business in Europe.

So this was the magnitude of the challenge Brian Chesky was suddenly facing in 2011 when he learned of Wimdu that spring. A few weeks after noticing the new rival, he got a phone call from Guy Oseary, a co-investor with Ashton Kutcher in Airbnb and a talent manager for musicians such as Madonna and U2. Oseary told Chesky that Oliver Samwer wanted to talk and soberly suggested that Airbnb might have to do a deal.

Chesky called Samwer, who seemed nonchalant on the phone and insisted that he was eager to build Wimdu himself. Nevertheless, he offered to fly to San Francisco right away so they could meet in person.

It was all happening with disorienting speed. Chesky, Gebbia, and Blecharczyk, along with investors Greg McAdoo and Reid Hoffman, met Oliver Samwer a few days later in the offices of the law firm Fenwick and West. (The Airbnb office on Tenth Street was too unimposing, and the so-called boardroom wasn't soundproof.) Chesky was stunned to see Samwer arriving directly from the airport carrying only a laptop bag. "I remember thinking that I had never seen a person leave a country without a change of clothes," he says.

Samwer confidently showed off the Wimdu website, a sister site called Airizu for the Chinese market, and an aggressive plan to hire four hundred employees and managers in countries around the world. Airbnb, whose founders still interviewed and discussed each potential employee to painstakingly measure for "culture fit," had only some twenty workers in San Francisco and another few dozen scattered around the world handling customer service and working mostly from their homes. Samwer suggested that Airbnb and Wimdu "partner." But the subtext was clear: he was holding the loaded gun of competition to Airbnb's head. The ransom was a merger. "We all just kind of looked at each other and said, 'Uh-oh,'" Chesky says. "It was pretty impressive."

After the meeting and a coffee with Samwer at Starbucks, the founders sat down alone to discuss the possibility of doing the deal. Chesky asked for Gebbia's and Blecharczyk's opinions; he was looking for consensus, but they were torn. They knew that a truly dominant home-sharing service would have to be global, offering the largest variety of lodging options to travelers, wherever they wanted to roam. They also knew that Samwer didn't share their values, their design sensibilities, or their desire to build a close-knit community. Samwer was so hard core and ruthless, the founders privately nicknamed him "the General."

To gather more information about their formidable adversary, a few weeks later the founders and board member Greg McAdoo flew to Berlin to visit Wimdu's offices. Chesky was astounded by what he saw inside a rehabbed factory in the city's Mitte District: Row after row of employees, most in their early twenties, sitting shoulder to shoulder at desks in the sweltering heat. There were no fans; it was literally a sweatshop.

Samwer gave the founders a tour. On many PCs they saw both the Wimdu and Airbnb websites in adjacent web browsers. "This is what we do," he told them unapologetically. "You Americans innovate. Me and my army of ants, we go fast and build great operations." He also told them that Wimdu had raised $90 million from Rocket Internet and other European venture capital firms and was already nine times larger than Airbnb in Germany.[11]

The founders and McAdoo went to dinner afterward and then stayed up all night in their nearby Airbnb, debating their options. Again Chesky pursued an elusive consensus among his colleagues, and reassurance that everything would turn out all right. But they were in a bind. They couldn't partner with the Samwers and stay true to their own values, and they couldn't fight the brothers without going on a rapid hiring spree in Europe to build a local operation and embracing some of the same ferocity. And they couldn't sit by idly either. "We are going to leave Berlin doing something that we weren't planning on doing before we started down on this journey," McAdoo recalls telling them.

One option was to find a local leader who could quickly spin up a European business to counter Wimdu and the other clones. The next day, they met one candidate at a café in the Berlin airport: a German entrepreneur named Oliver Jung, who had been referred to them by talent manager Guy Oseary.

Jung was tall and bespectacled and had the same spotty history of cloning companies as the Samwers. Over the previous few years, he had backed a LinkedIn knockoff called Xing; a members-only shopping site called Beyond the Rack, which looked remarkably like Gilt; and a Swiss deals site called DeinDeal. He was also just as intense as the General. When the founders told him about their predicament, he started pacing back and forth in the coffee shop, a Bluetooth earpiece in his ear, and ordering colleagues in Barcelona, Paris, and elsewhere to get on planes and begin coordinating a response. He knew the Samwers well—they had invested together in several startups. "I knew how insane Oliver Samwer is," Jung says. "I had so much respect for him that I was scared."

By the time the founders returned to San Francisco, Chesky was reasonably certain that Airbnb was not going to work with the Samwers. But he didn't yet know whether they could work with Oliver Jung, who seemed just as mercenary. It was also unclear who at Airbnb would lead the response to the company's greatest challenge yet. Over dinner with Chesky at a Thai restaurant that week, McAdoo was adamant—it had to be Chesky himself. Even though the twenty-nine-year-old CEO had barely traveled outside the country, and even though he knew next to nothing about how to build a large global organization, how to manage a company, or, frankly, how to make decisions without pursuing time-consuming unanimity among his co-founders, Chesky would have to rise to the occasion and finally embrace the responsibility of his job title.

"You among the founding team are the only one with the intuitive sense of what this needs to look like and the drive and passion to be able to lead it," McAdoo told him. "So, yes, let's hire Oliver Jung, and, yes, we have to bring in senior people. But you will spend a lot of time on airplanes learning how to build and run a large-scale global organization. Are you up for that challenge?"

Chesky had barely had a chance to answer the question before, a few weeks later, a completely new and unrelated crisis erupted.

> Three difficult days ago, I returned home from an exhausting week of business travel to an apartment that I no longer recognized. To an apartment that had been ransacked.

A host using only the initials EJ had written on her WordPress blog that her San Francisco home had been burglarized and trashed by a guest who had rented it for a week via Airbnb.[12]

> They smashed a hole through a locked closet door, and found the passport, cash, credit card and grandmother's jewelry I had hidden inside... They rifled through all my drawers, wore my shoes and clothes, and left my clothing crumpled up in a pile of wet, mildewing towels on the closet floor... Despite the heat wave, they used my fireplace and multiple Duraflame logs to reduce mounds of stuff (my stuff??) to ash... The kitchen was a disaster—the sink piled high with filthy dishes, pots and pans burnt out and ruined... The death-like smell emanating from the bathroom was frightening.[13]

EJ had lost everything.

She tore into Airbnb for presenting the illusion of mutual trust between hosts and guests when apparently its faith in humanity was grossly misplaced. She faintly praised the company for eventually responding to her entreaties, writing that the customer-service team had been "wonderful, giving this crime their full attention." But she would soon change her tune.

EJ's blog post remained largely unnoticed for a month. Then, in late July, after the financing from Andreessen Horowitz had been made public, the incident became a hot topic of discussion on Hacker News, a popular online bulletin-board site run by Airbnb's first benefactor, Y Combinator. The site's users offered their own thoughts about the incident and began a lengthy debate over the honesty of the common man.[14] Then, Michael Arrington, the imperious founder and chief blogger of *TechCrunch,* noticed the thread and wrote an article about the incident titled "The Moment of Truth for Airbnb As User's Home Is Utterly Trashed."[15]

Arrington spoke with Chesky for the article, and Chesky told him that the company knew about the incident and had offered to financially assist EJ, help her find new housing, and do "anything else she can think of to make her life easier." Moving swiftly to further contain the damage, Chesky then wrote his own article for the tech-news site, emphasizing that the Airbnb execs were "devastated" by the incident but had been "in close contact" with EJ since the beginning.

That's when the shit really hit the fan.

The day after Chesky's article appeared, EJ returned to her blog, full of fury. It turned out that Airbnb had not, in fact, sent the promised compensation or come through with any alternative accommodations. (According to multiple, conflicting conversations I had with current and former employees from the time, the payment was either never authorized or never sent.) The primary person she had talked to from Airbnb was the brainy Blecharczyk, who was filling in for a recently departed head of customer service while Chesky interviewed numerous potential replacements. Blecharczyk, EJ wrote, spoke of "his concerns about my blog post, and the potentially negative impact it could have on his company's growth and current round of funding." The CTO then suggested, EJ wrote, that she shut down her blog or update it with a "twist" of good news.[16]

EJ described herself as basically homeless, terrified, and "broken" by the situation. Some readers offered to send her money, but she told them to keep it and "book yourself into a nice, safe hotel room the next time you travel."[17]

It was a devastating rebuke. For the next five days, a nonstop Twitter-fueled media storm drew Airbnb into its torrential winds and refused to subside. The tech-news sites piled on—Airbnb was bringing strangers together in homes without ensuring a safe experience. If this host's home had been vulnerable to such methodical destruction, what could be in store for other people? Commenters on blogs suggested protesting at the founders' residences; #Ransackgate became a trending topic on Twitter, and the story was covered by major outlets such as CNN, *USA Today,* and the *San Francisco Chronicle.* Chesky, Gebbia, and Blecharczyk had given Silicon Valley critics a new reason to think tech startups were just as rapacious and careless as every billion-dollar business before them.

Just a year ago, the startup had consisted of its three founders and a handful of employees sitting around a table inside the apartment on Rausch Street. "We were getting treated like an adult but we hadn't truly grown up yet," Chesky says. But that was an inadequate excuse and he knew it. Every savvy investor that he had ever pitched had raised burglaries or other crimes as one possible result of home-sharing. Yet Airbnb hadn't been ready and had made negligent mistakes that were unacceptable for a company recently valued at $1.3 billion. "There were a lot of good questions, like how in the hell could a billion-dollar company not have its shit together," Chesky says.

EJ had also raised fundamental questions about the safety of users on its site and Airbnb's role as an arbiter between hosts and guests. Until that incident, Chesky had subscribed to the purist's view of online marketplaces: Users were supposed to police one another by rating their experiences. Untrustworthy actors would be drummed off the platform by bad reviews, rejected by the web's natural immune system.

It was a libertarian view of the internet and had the whiff of Silicon Valley snake oil. The prospect of a negative review is of little use after a serious breach of etiquette—or a criminal act. But because of their shared faith in the power of self-policing marketplaces, Chesky and his colleagues hadn't made serious investments in customer service or customer safety. The fact that Blecharczyk, as well as the company's controller, Stanley Kong, had been put in charge of customer service at a company now with over 130 employees while the other founders looked for an executive to run the department was telling. "We viewed ourselves as a product and technology company, and customer support didn't feel like product and tech," Chesky says.

The three founders describe the week following EJ's second blog post as the most difficult of their careers. They had told everyone, and themselves, that Airbnb was bringing people together and making the world a better place. Now the company had abetted a serious crime and bungled the fallout. One night during the madness, the founders drove forty-five minutes south to the home of their first mentor, Paul Graham. He had never seen them more forlorn. "They just want to see you suffer," Graham told them in his kitchen. "They want an ounce of blood. Just fall on the sword, accept responsibility, and everyone will move on."

Over the next few days, Chesky drew his investors and advisers close around him and fashioned a suitably contrite response. The company would introduce a twenty-four-hour customer-service hotline, double the size of its customer-support staff, and form an in-house trust and safety department, separate from customer service, that focused on fighting fraud and addressing poor experiences on the service. The startup also began working on ways to verify users' identities; for example, by making it clearer when customers had manually confirmed their phone numbers or connected their Facebook accounts to Airbnb.

The plan's centerpiece offering was dubbed the Airbnb Guarantee. Jeff Jordan, the partner at Andreessen Horowitz who had joined Airbnb's board of directors, had introduced a similar program at eBay, called Buyer Protection, that adjudicated disputes between buyers and sellers and gave refunds to aggrieved customers. Jordan suggested it could work here as well. Chesky intended to set the guarantee at a modest five thousand dollars. Then Marc Andreessen visited the Airbnb office one night to support the beleaguered founders and suggested they should add a zero to their announcement and reimburse hosts for up to fifty thousand dollars. It was a significant risk at the time, since the company didn't have insurance and would have to cover any costs itself.

Airbnb was effectively betting its enormous haul of venture capital on the premise that tragedies like the one that had beset EJ would be rare. (The following year the Airbnb Guarantee would grow to a million dollars, insured by Lloyd's of London.)[18] "There was an element of Butch and Sundance jumping off the cliff at that point," says Jeff Jordan. "They had this belief that people are basically good and that almost all trips were positive trips."

Airbnb also hired Brunswick, a crisis communications firm, which recommended that Chesky write a letter to his customers. They proposed a draft, but Chesky thought it sounded jargonish and evasive. He felt beset by conflicting advice and unsure of his own instincts, since what he had done previously had only made things worse. Eventually, he decided to speak frankly to his customers himself, and he rewrote the letter with Ligaya Tichy, an early Airbnb marketing exec.

In the spirit of Graham's advice, Chesky impaled himself fully on his sword. "Over the last four weeks, we have really screwed things up," he wrote. "I hope this can be a valuable lesson to other businesses about what *not to do* in a time of crisis, and why you should always uphold your values and trust your instincts."[19] He appended his personal e-mail address to the letter—another recommendation from Andreessen.

Chesky talked to his board of directors over the phone that weekend and announced these moves. On the morning of August 1, the letter was e-mailed to Airbnb's one million users and widely dissected by the press. As Paul Graham had predicted, the internet storm was quelled; the mob moved on. Some observers expressed disappointment at this, since watching a newly anointed startup crash and burn was heady entertainment.

The EJ saga receded from public view but still played out behind closed doors. EJ's destructive guest, nineteen-year-old Faith Clifton, was arrested that summer in San Francisco on charges of possession of stolen property and methamphetamine, fraud, and an outstanding warrant in a nearby city.[20]

EJ, an event planner in her late thirties named Emily, continued to press her case against the company. According to a former Airbnb employee, the parties entered into mediation that year and Airbnb agreed to pay her a hefty settlement, sealing the entire episode up with a nondisclosure agreement. She later declined to talk to me, writing in an e-mail, "That is a chapter in my life I have long since buried and do not wish to revisit." Chesky and Airbnb also declined to discuss the outcome of the EJ affair.

The company had quieted the furor, survived one of its greatest challenges, and added new protections for its customers. But in the end, it couldn't change human nature. The private payment to EJ, says the former Airbnb employee, was among the first of many the company would make to customers whose experiences went horribly, and sometimes tragically, awry.

That same summer, as if the whole wrenching saga hadn't even happened, Airbnb moved to a new office at 99 Rhode Island Street, at the foot of San Francisco's tony Potrero Hill District. For the first time, Chesky and Gebbia got to stamp their design sensibilities on their workspace. There were long stylish desks, Eames chairs, beanbags, a tree house where employees could take naps, and an antelope head on the wall in the bathroom. Three conference rooms were modeled after rooms available for rent on the site, and inspirational sayings like *Life is lovely* were printed on the walls.[21]

In August, Airbnb had a party to celebrate the new digs. MC Hammer DJ'd on the rooftop while guests danced, played Skee Ball, and sipped cocktails. At one point the founders stood on chairs and addressed the crowd. Joe Gebbia wore a white tuxedo shirt with white and blue ruffles down the front and a panama hat. The technology press uniformly interpreted the exuberant affair as evidence of another doomed tech bubble. But this was only the summer of 2011. They hadn't seen anything yet.[22]

Despite the playful façade, Airbnb was still at war. Oliver Jung didn't hear from Brian Chesky for a month after the whirlwind meeting at the Berlin airport. There was a good reason—Chesky was consumed with the EJ saga. Over those weeks, while Jung grew more pessimistic about his chances of working with the startup, he got more enthusiastic about its prospects. By chance, he had coffee with an old friend from Madrid who had rented out his apartment through Airbnb for six months and financed his travels with the proceeds. Airbnb hadn't even advertised to lure the friend to the service—he had just seen news articles. Jung saw the possibility of a business with global reach and little in the way of customer-acquisition costs.

Finally, in late summer, Jung called Chesky to check in, and Chesky delivered the news: He had decided not to partner with the Samwer brothers and their clone Wimdu. Summoning up his resolve, Chesky had told his co-founders, employees, and investors that "he would rather not negotiate with a terrorist and go fight and lose than give in," according to Alfred Lin, a partner at Sequoia. In a curt phone call, Chesky had bluntly informed a stoic and nearly wordless Oliver Samwer of his decision. Chesky was now ready to coordinate the response, and he invited Jung to come to America to discuss it.

Jung flew to San Francisco the next day and was thunderstruck when he arrived at the Rhode Island Street office and witnessed its quirky rituals, like lunchtime yoga and a weekly company-wide kickball game. He had heard about the frenetic offices of Wimdu and its "army of ants." This was the opposite. "It felt to me like only thirty people were there and everyone was very relaxed," Jung says. "Some people were playing table tennis. Then someone brought a dog out, and it was its birthday. Everyone celebrated the birthday of the dog."

Jung's initial reaction was panic. *Oh my God, Wimdu is going to kill them,* he thought. Then Chesky came out to greet him and introduced him to every employee at the company. Jung spent the day in a succession of interviews, eventually getting on the phone with McAdoo, who peppered him with questions about who he would hire as country managers and how he would build a global team. By early evening, Chesky and his partners seemed satisfied. When Jung signed a contract to make a personal investment in Airbnb and serve as the head of its international expansion, Chesky told him, "This is going to be the best deal of your life." Jung had already made millions investing in startups in Europe and Israel, but Chesky was right, by an order of magnitude.

Chesky wrote the business plan for the international expansion. Each new regional office that Oliver Jung set up would take charge of cultivating the supply of rental properties and supporting the community of hosts. The San Francisco team would produce the underlying technology and coordinate marketing and publicity to generate demand. The goal was to export the things Wimdu didn't have and didn't care to replicate—Airbnb's sense of mission and the way it cultivated an intimate community among its users. Chesky assigned Lisa Dubost, one of his earliest employees, to work with Jung and chose Martin Reiter, a new head of international operations, to vet the new hires and ensure all the new country leaders embodied Airbnb's corporate values and fit the profile of its employees.

That fall, after putting pins on a wall map and considering the best way to halt the Samwers' momentum, Jung opened new offices in Berlin, London, Barcelona, Copenhagen, Milan, Moscow, Paris, Delhi, India, and São Paulo, Brazil. In June, the company bought one of the smaller German clones, Accoleo, and opened an office in Hamburg.[23] Jung traveled throughout Europe and Asia and had dozens of interviews each day with prospective country managers; it was like speed-dating. At the end of every interview he asked the candidate: "How do you feel [about it]?" If the individual was energized by the opportunity and likable, Jung sent him or her to San Francisco, where Chesky had the final say.

Airbnb furnished each new manager with a set of online tools to monitor the health of the business and with something Chesky called an "office in a box." It contained a guidebook to setting up an Airbnb-like working environment and included various props, like a portable Ping-Pong table and the books *Delivering Happiness* by Zappos founder Tony Hsieh and *Oh, the Places You'll Go* by Dr. Seuss. "Brian was always worried about how do we scale our culture—how does every Airbnb office feel?" says Dubost, who became Airbnb's vice president of business travel and left the company in 2016.

Some of the new regional managers modeled themselves after Chesky. In Moscow, Jung hired Eugen Miropolski, a former Groupon exec, who promptly rented out his home and started living in Airbnbs around town, just as Chesky had done in San Francisco. In Paris, Olivier Grémillon, a former McKinsey and Company consultant, planned community meet-ups to greet hosts and instituted a 24/7 customer-support line manned by French speakers so hosts and guests always had someone to call.

In January 2012, Airbnb publicly announced the opening of its international offices. The three founders hit the road, each attending a launch party in a different city and coming together for blowouts in Paris and Berlin. Chesky recalls hardly sleeping for eighteen days.

They trained their new employees, gave speeches about the warmth and potential of the Airbnb community, met hundreds of hosts, and doled out countless hugs. "It gave you a feeling that it was not business oriented," said Nalin Jha, one of the earliest hosts in Delhi, India, who joined the service that year after attending the company's first local meet-up and recalls being immediately embraced by the general manager Jung had hired. "It was just a small hug, but it suggested there was a soul in the business. That was a very attractive thing, that I was becoming part of a community."

Oliver Jung figured that the Samwers had a year's head start outside the United States. But Wimdu didn't last. Like the brothers' Groupon clone, Wimdu was a hollow company whose momentum was based on a deluge of impersonal sales calls—not on meet-ups and certainly not on hugs. Airbnb had more robust technology tools, furnished by Blecharczyk and his team of engineers in San Francisco, and the benefit of a global network. U.S. travelers to Europe didn't seem to care about Wimdu's early dominance, and European travelers to the United States who sought alternative accommodations had to turn to Airbnb.

Wimdu stuck around but would become an inconsequential player in the home-sharing market. In 2013 it shuttered Airizu, the Chinese subsidiary, and pared back its ambitions outside Europe. Airbnb had shown Silicon Valley that it was better to fight cloners than to accommodate them. "The worst thing you can do to a cloner is to let him keep his baby," Chesky joked to Oliver Jung. "The cloner doesn't want his baby. They build the baby to get rid of it." Meanwhile Airbnb continued to gain altitude. In January 2012 it announced that it had booked a cumulative five million nights since opening for business, and by June it had already updated that figure to ten million.[24]

The international expansion was a success. That year Jung added offices in Singapore and Hong Kong, and by the end of 2012, Europe already was Airbnb's largest market and Paris its largest city.

Yet the effort wasn't perfect. There was high turnover in the new offices, and Jung ended up quitting the company in early 2013. Some of the offices were consolidated; some leaned too heavily on hosts with multiple listings, which the company was trying to discourage in favor of people who were sharing their own primary homes. And in the U.S. headquarters, the rapid overseas expansion generated waves of anxiety. Employees of the small company, where everyone knew everyone else, were offended by the idea that there were now hundreds of new colleagues around the world that they had never met. "Everyone seemed to know everything that was going on and suddenly they didn't know anything," Chesky recalls. "It was very controversial. People did not like it."

Chesky wasn't happy about the internal discord but was learning to live with it. Amid the turmoil of 2011, he had found his footing and embraced his job as the company's top decision maker. He had laid out a course of action to deal with the EJ mess and chosen to battle the Samwers rather than take the easier path of working with them. He still listened to colleagues and his co-founders, but after that year, he no longer tried to seek consensus. Instead, he surveyed opinions and trusted his own instincts to make a decision.

"This is when I became CEO in a meaningful way," he told me years later. "I changed my style. I hope *jerk* isn't the first word they use. But 2011 was the year I really had to become a CEO, to become a champion of Airbnb, to get people to want to believe in it, to raise money, and to get us out of a true crisis. We came out of EJ and the Samwers stronger than when we came into it."

With cities waking up to the problems posed by people turning their homes into ad hoc hotels, Chesky's mettle would soon be tested further. He would have to prove to suspicious lawmakers and regulators that Airbnb's intentions were pure and that its impact on cities was constructive. It would be his most serious challenge yet and one that Travis Kalanick, Chesky's new friend and a peer in the proliferating movement called the sharing economy, was about to face in an even more potent form.

CHAPTER 7

THE PLAYBOOK

Uber's Expansion Begins

> I've never seen an entrepreneur work as hard. He lives, eats and breathes Uber.
>
> —Shervin Pishevar, e-mail to his partners at Menlo Ventures

To Travis Kalanick, Uber wasn't merely a fecund investment opportunity or a promising startup with an auspicious set of early results. As he described it at the start of 2011 to friends and colleagues, the company was a blossoming passion—the entrepreneurial jewel that he had coveted his whole career.

Kalanick was ready to devote all his energies to the new object of his affection and he expected his employees to work just as hard. And he aggressively confronted and expelled from his inner circle anyone he thought might stand in the way of Uber's manifest destiny—to grow beyond the enclave of San Francisco and conquer the world.

Unlike Airbnb, Uber had a lot of work ahead to spread internationally. Airbnb had instantly become global upon its launch. Motivated by competition from the Samwers, Chesky and his colleagues had attacked that opportunity head-on. Uber, however, was going to have to methodically enter each market and find employees in every city who could recruit drivers, promote the service to riders, and talk to regulators. Compared to Chesky's different and somewhat easier path, Kalanick's effort to build a global empire was going to be laborious.

Kalanick's first target was New York City, home to half of all cab rides in the United States. Unlike sprawling Los Angeles, his hometown, with its unrepentant car culture and freeway gridlock, the Big Apple was the country's densest metropolitan area, where most residents avoided owning a car. If Uber could make it there, perhaps it could make it anywhere.

To lead its New York expansion, Uber hired a fresh-faced Cornell University graduate named Matthew Kochman. As a college junior, Kochman had started a campus charter-bus operation called MESS (Moving Every Student Safely) Express, a service that let sororities and fraternities book rides online, reducing the likelihood of drunk driving. The business took off, and on many trips, Kochman, tall and good-looking, would sit at the front of the bus and talk on the intercom, entertaining his fellow students.

After graduation, Kochman moved to New York to start a company that would allow parents to text cab-fare money to their kids. He tested the service with a taxi company in Ithaca and contracted with a development group in Uganda to build the service. But the Ugandans didn't come through, and just as he was starting to doubt whether the business would work, he attended a technology conference in San Francisco and read an article about Uber. He e-mailed Ryan Graves and met him for coffee, and Graves was evidently impressed by his experience and youthful hustle. A few weeks later, Graves e-mailed him to see if he was interested in becoming Uber's first general manager outside of San Francisco.

Kochman opened the first Uber office in a coworking space on the corner of Broadway and Grand Street in lower Manhattan and spent the initial few months scrambling to establish Uber in New York City. The pitch that had worked so well in San Francisco on town-car drivers—make money instead of sitting around waiting for passengers—clanked pitifully off the polished black hoods of New York City town cars. There was an arcane but important regulatory reason for this. According to the city's byzantine taxi codes, drivers of town cars (livery cars, in regulatory parlance) had to be affiliated with some kind of base, either a professional fleet or a small local organization that acted as a central dispatcher and assigned rides while ensuring that cars were properly licensed. But Kalanick refused to register Uber as a base. He believed that it would make the company responsible for various fees and licensing requirements not only in New York but in other cities where it wanted to expand, and he felt that Uber should stay free of the regulatory muck. Though the base law was rarely enforced, Uber was technically trying to entice drivers to break it by signing them up for a secondary source of rides.

By April, Kochman had found a few adventurous livery drivers looking to fill the dead time between rides and was testing the service across all five boroughs.[1] The service quietly launched the next month at a meet-up, a gathering of the local tech community, but there were only a few cars on the road. Kochman was under a mountain of pressure. Kalanick wanted to unveil Uber's second city to the broader public at TechCrunch Disrupt in June, where he was set to make that joint appearance with Brian Chesky. Kochman hired two employees, one to oversee driver operations and another to promote the service to riders. To spark the business, they started offering the same perks that had worked in San Francisco: drivers received iPhones with the Uber application and were guaranteed a minimum twenty-five to thirty-five dollars an hour. Passengers who tried the app but didn't see any cars were given ten-dollar credits. It wasn't long before Uber was hemorrhaging money.

With a limited number of cars, Uber wait times in New York were unacceptably long. On the day of Disrupt, Kochman recalls, Uber had about a hundred cars on city streets. (To put that in perspective, Uber had more than thirty-five thousand active drivers in New York City in 2016.)[2] When customers in the world's largest media market signed onto the app, they were either seeing no cars available ("zeros") or waiting more than ten minutes for a ride. Kalanick wasn't happy. "When demand outstrips the quality of service, pickup times and other items are not quite where we want them to be," he said onstage.

The next few months of Kochman's life were a frenzy of stress and around-the-clock work. "We need more fucking cars!" Kochman recalls Kalanick barking over the phone.

"You've just got to hustle more," Graves reassured him unhelpfully. "You've got to grind."

Kochman and Kalanick didn't see eye to eye on many of Uber's pressing issues in New York. Still embittered by his experience with Christiane Hayashi and the SFMTA, Kalanick instructed Kochman to ignore New York's Taxi and Limousine Commission and its rules, reasoning that its regulations, under the guise of consumer safety, were really there to protect entrenched taxi interests.

Kochman didn't necessarily disagree but he had successfully solicited approval from the Ithaca city council to operate MESS Express. He had a history of productive encounters with regulators. So he disregarded Kalanick's orders and set up a meeting with a TLC deputy. "I wasn't going to bust my ass to launch something the city was going to shut down immediately," he says.

Kalanick was furious when he heard about the meeting. "He was absolutely livid and said it was insubordinate," Kochman remembers. After his fury passed, Kalanick flew to New York and they visited TLC headquarters together. The meeting—Uber's first of many with the TLC—went fine. Stressing that Uber cars were not hailed or even electronically hailed like taxis, the pair emphasized that Uber cars fit the legal definition of livery cars and were prearranged; it just so happened that the prearrangement occurred five minutes ahead of time instead of sixty. The deputy commissioner Ashwini Chhabra, who would join Uber as head of policy planning three years later, initially asked Uber only to customize the app to display a driver's permit number and base affiliation.

Uber may have had nodding approval from a regulator, but it still didn't have enough drivers. In search of a silver-bullet solution, Kochman visited midsize limousine and town-car fleets, just as Kalanick and Graves had originally done in San Francisco. One day he ventured into Brooklyn, to an office a block away from the Gowanus Canal, to meet with Eduard Slinin, the Ukrainian founder of the Corporate Transportation Group, an umbrella organization of a dozen or so livery fleets. If Kochman could enlist the help of Slinin and his thousands of cars, he could solve all of Uber's supply woes with a single handshake.

He spent two hours pitching Uber to Slinin and seven of his stony-faced, pinstripe-suited colleagues. Then he listened to a barrage of reasons why Uber would never work in New York City: regulators would object, drivers were too busy to consult smartphones, and the big banks and law firms already had relationships with limo fleets. "Listen, I like you," Slinin told him at the end of the conversation, according to Kochman. "But I would advise you not to launch Uber in New York. It would not be good for you."

Kochman left the meeting upset, interpreting that as a physical threat. (In an interview, Slinin denied ever threatening Kochman.) When he heard about the incident, Kalanick was untroubled. "If you get whacked, do you have any idea how much press that will get us?" he joked. Kochman was annoyed. Like Jason Finger at Seamless years before, he wondered whether he was going to live in mortal fear of every livery car that passed him on the street.

Kochman felt that Kalanick's spirited combativeness might be working against Uber. He got far in negotiations with one fleet, the Executive Transportation Group, or ETG, which ran about two thousand town cars in the city. But when he brought Kalanick to a meeting with its executives, they grew suspicious. For good reason; after the meeting, Kalanick turned to Kochman in the backseat of an Uber and said: "We are going to stab those guys in the back." Kalanick later remembered this differently, saying he was genuinely in "partner mode" back then.

Regardless, Uber's alliances with the large limo fleets were, in fact, fated to fail. The company would ultimately challenge them by providing drivers with a steady source of riders, allowing those drivers to save on the significant cut they had to hand over to the fleet owners like CTG and ETG.

Kochman remembers something else his boss said on the ride back from the ETG meeting. Kalanick was talking with awe and envy about Jack Dorsey, who was fired from Twitter in 2008 but had recently reemerged with a more refined image at the payments firm Square. Success on the internet, it seemed, could be a platform for personal reinvention—a way to rid oneself of all the baggage of the past.

"I remember on that car ride, Travis was explaining to me that Jack was a very different person early on in his career," Kochman says. "After Twitter, he went away, he disappeared off the map and self-reflected. And then he came back as a completely different person."

By the spring, the relationship between Kalanick and Kochman was deteriorating rapidly. Kalanick wanted to see faster growth in New York to show the venture capitalists in order to raise additional capital and finance an expansion into other major U.S. cities.

Meanwhile, Kochman viewed his boss as disruptive, coming to town to meet with investors and walking into the lower Manhattan office with dreams of seemingly irrelevant future services, like Uber cars that could deliver meals. Accustomed to being in the spotlight at college, Kochman also privately fumed that it was Travis, always Travis, at the center of media attention.

Then things fell apart entirely. Kochman believed that as Uber's first general manager outside San Francisco, he had a substantial ownership stake in the company. But while hiring new employees in New York, he discovered that his portion of stock had been allocated *before* the Series A funding round from Benchmark and Bill

Gurley, not *after,* as he had originally believed. This meant that his ownership percentage was significantly smaller than he had thought, since new investors in a startup dilute the equity stakes of older shareholders. Kochman thought he had been deliberately misled and was furious. During a tense discussion with Kalanick at the Mondrian Hotel in SoHo, he suggested that Kalanick had been opaque and disingenuous during their negotiations over his compensation.

Kalanick wasn't in the mood to hear it. "You're an employee. We pay you a salary. Do your fucking job!" he said.

Kochman then concocted a dubious plan. He sent an e-mail to Bill Trenchard, a partner at First Round Capital whom he knew from his years at Cornell, that outlined a litany of complaints against Kalanick and Graves. He wrote that there was a widespread lack of trust and confidence among the staff in Uber's leadership, that at least five key employees were contemplating leaving, and that "management is horrible at listening." At the end of the note, he suggested a "rearranging of management" and asked Trenchard to circulate the memo among other Uber investors.

Nothing happened, so a few weeks later, the still angry Kochman invited Kalanick out for lunch again. Though Kalanick apologized for the tenor of their last conversation, Kochman announced his intention to quit. He gave three months' notice and left in September, without staying a full year or collecting any portion of his fifty thousand shares of stock. What he couldn't have known was this: in just a few years, those shares would be worth more than a hundred million dollars.

When I met Kochman in early 2015, he was still enraged by Kalanick's treatment of him and Uber's sharp-edged business tactics with competitors and drivers. But when we met for a second time a few months later, to my surprise, his fury had subsided and his tone had softened. He was finally coming to terms with his own ghastly, youthful mistake.

"In my twenty-three-year-old head, I concocted a game plan that would ultimately lead to Travis being ousted and me put in his place. That was my legitimate intention," he told me at a café in Williamsburg, near the small office of his new charter-bus startup, Buster (it would go out of business soon afterward). Following his departure from Uber, he said, he talked to the media about problems at the company, advised prospective employees not to take jobs there, and counseled venture capital firms not to invest. He also consulted for both Lyft and Hailo, a UK-based taxi-hailing app.[3]

"I've definitely done my fair share of hyperbolic shit talking," Kochman said. "At the end of the day, even though we disagree on a bunch of things, Travis Kalanick is a brilliant guy who built a massive business and I'm proud to have been a part of it."

He has tried unsuccessfully to get in touch with Kalanick but doesn't blame him for not responding. Recently, he says, he had a vivid dream about hashing over the past with him at Citi Field during a Mets game. "But that will never happen." Kochman sighs. "He hates me."

Austin Geidt had been adrift when she started working at Uber. The company's first intern got the job after being rejected for a barista position, and she struggled to find a meaningful role. But during that first difficult year, she had a dawning realization: just about everyone else around her was making things up as they went along too.

After the epiphany, Geidt gave herself permission to look at problems more constructively. When Ryan Graves made her head of driver operations that March after firing another early employee, Stefan Schmeisser, she had ample opportunity to make a difference. After training a driver in the office in San Francisco one day, she walked outside to get coffee and saw the man getting into a pink minivan. It occurred to her that the company should probably conduct vehicle checks to ensure cars were up to Uber's high standards at the time.

Later, she decided Uber should test whether drivers had at least a cursory knowledge of city landmarks (this was before Uber started having riders enter their destinations into the app). She asked Sofiane Ouali, the driver of white 2003 Lincoln Town Car known as the unicorn, to obtain for her the city's entrance exam for yellow-cab drivers. Then, with help from Ouali and other drivers, she changed some of the questions to match the expectations of Uber's upscale, smartphone-using clientele. Instead of asking directions to the city jail, for example, Uber would ask drivers if they knew the location of the Ritz-Carlton.

Geidt worked closely over the phone with Kochman in New York that year and was the sounding board for his stream of complaints about Kalanick. In July, when Kalanick selected Uber's third city, Seattle, and started Uber's furious national expansion in earnest, he dispatched Geidt and Graves to open the new office and to hire the launch team. She had just signed a lease on an apartment in San Francisco but wouldn't spend a single night there. The next year and a half, she lived on the road.

Geidt and Graves based the Seattle operation on the three-person structure in New York. The general manager supervised the overall business in the city and was accountable for its growth. He or she needed to be entrepreneurial, scrappy, and aggressive in talks with regulators. An operations manager, usually an analytical type like a management consultant or investment banker, was in charge of signing up drivers and making sure there were cars for every passenger who opened the app. Finally, a community manager, a creative type with marketing chops, worked to stimulate demand among riders.

This became the early template for local offices, Uber's equivalent of a SWAT team, able to drop into cities and rapidly spin up a new business. "This is totally unique for a tech company," Kalanick told me in an interview during the initial expansion of the black-car service. "Technology companies used to be all product and engineering, sitting at its headquarters. When you scale up, you just light up another machine. When we scale we have to get more cars on the ground and make sure we are onboarding drivers who will provide a quality experience."

Along with Kalanick and Graves, Geidt pioneered other aspects of the model and recorded it all in an online Google document to serve as a manual for Uber's entrance into new cities. Drivers should be solicited by combing through limo-fleet listings in Yelp, the online directory, or by visiting airport-limo waiting lots. A launch party should bring together local media and tech luminaries, while a local celebrity should be selected as the first rider in the city and promoted in a blog post.

They also used strategies to attract both drivers and riders, like offering subsidies and credits, and took some basic but important steps, like opening an Uber Twitter account in each city. The Google doc would become a company bible; employees took to calling it "the playbook." Seattle, says Geidt, "became the first iteration of our playbook."

Geidt spent the next few weeks there, skipping the September launch of Chicago, Uber's fourth market, but moving to open a fifth city, Boston, a few weeks later. For a company that had gestated quietly for three years, it was all happening with lightning speed. Back at the home office, now half a dozen desks at the downtown office-sharing firm RocketSpace, Kalanick was tracking daily results from each market and comparing them to the early patterns from San Francisco. Every GM was responsible for staying above the original trend line. Bill Gurley, observing from the sidelines, was impressed. "I've watched hundreds of entrepreneurs go to other cities fast and wreck the whole ship," he says. "I never had that anxiety with Uber. This was systematic. There was a massive amount of math used in the decision-making process."

A day after the Boston launch, Geidt got a call from Graves—the company wanted her in New York City. Matthew Kochman had left and his deputies were following him out the door. Uber desperately needed help in the largest and most important taxi market in the country. Not only was driver growth still slow but the TLC was being inundated by complaints about Uber from taxi and limo drivers, just as the MTA had been in San Francisco. Officials were now expressing concerns about Uber's regulatory compliance and threatening the company with a cease-and-desist.

Sick of staying in hotels, Geidt checked into an Airbnb in the East Village. She would live there for five months. Right away she found that getting an Uber from her place to the newly opened New York office in Greenpoint, Brooklyn, was a hit-or-miss affair. Cars were scarce and wait times were high. The large livery fleets had all the leverage and demanded outrageous minimum payments to give the Uber app to their drivers. Uber had to rethink everything about its tactics. And Kalanick would have to compromise on some of his most stubbornly held convictions.

The first step was to start in-depth talks with the TLC. To lead the charge in New York and in what promised to be a swarm of regulatory challenges in every city, Kalanick hired his first lobbyist: Bradley Tusk, a former aide and campaign manager to then mayor Michael Bloomberg. Kalanick met Tusk at his office in midtown and asked about his retainer. Tusk told him it was twenty-five thousand dollars a month.

"Is that cash? That's a good business," Kalanick mused. "How about some equity instead?"

Tusk agreed and received fifty thousand shares, the precise amount that had just been abandoned by Matthew Kochman. It most likely was and remains the most lucrative lobbying contract in the history of that indecorous vocation.

After Tusk joined as a consultant, Uber executives started meeting regularly with Ashwini Chhabra and his boss, David Yassky, chairman of the TLC. Officials in Bloomberg's business-friendly administration, it turned out, were inclined to look favorably on a technology startup trying to change New York's crusty taxi industry, which had resisted modernizing its vehicles and installing electronic credit card readers.[4] But Uber first needed to play by the rules. To truly appeal to New York drivers, Uber was going to have to register as a base.

For all his feisty combativeness, Kalanick wasn't yet the consummate rule breaker of his later public image. He now saw that registering as a base in New York was in the company's best interests. According to the TLC bylaws, anyone with a 10 percent or greater ownership stake in the licensed entity had to be fingerprinted and sign the base application in person. So, on October 19, 2011, Garrett Camp, Bill Gurley, and Travis Kalanick all converged on the drab, fluorescent-lit TLC branch office and waited in line for an hour. "That was one of the first times where Gurley had to do something he thought was crazy," says Camp.

Registering as a base was only the first step in reconfiguring the New York strategy. Poring over the data from the first seven months in the city, Uber execs realized that their meager supply of drivers was spread too thin over the city's three hundred square miles. If they couldn't add drivers faster, perhaps they could redirect the ones they had to the busiest neighborhoods. So Geidt and the ad hoc New York team began directing drivers to spots where people were most likely to embrace an upscale thirty-five-dollar car ride, places like Wall Street, the Upper East Side, and SoHo.

Uber basically broke New York City into a series of targeted micro-cities. The execs called it "the SoHo strategy," and it would be a key element of the upcoming global blitz. Sending drivers to the places they were needed most would ensure a good experience among the social groups most likely to use Uber; they would then tell their friends, generating demand that would in turn make the service more lucrative to drivers. "We learned that you can't bring a San Francisco solution to New York and expect it to work," says Ryan Graves.

The SoHo strategy paid immediate dividends. Uber's engineers became adept at monitoring the service and identifying the hottest clusters of rider requests. Wait times went down, and the appeal of

Uber to New York drivers went up. And by registering as a base, Uber was no longer asking drivers to bend the law by accepting rides through a smartphone app. After these two critical changes, Uber's black-car service started to grow quickly in New York, just as it had in San Francisco.

That further emboldened the already driven Travis Kalanick. That fall, sensing he would have to rush to beat copycats to many of the world's largest cities, he asked his engineers to prepare the service for its sixth market: Paris. Once again seeking to take advantage of the attention that comes with an onstage appearance at an industry conference, Kalanick wanted to launch the company's first international city during LeWeb, the European technology confab where, three years before, he and Camp had hashed over plans for a hypothetical on-demand car service.

By then the startup had finally moved to its own office, on the seventh floor of 800 Market Street. It had a round conference room with broad windows that opened up onto Market, the city's main commercial artery. There were twenty employees in the new office, mostly engineers and data scientists, and another dozen in the field.

The engineers rebelled against the idea of opening overseas so soon. Launching in Paris required accepting foreign credit cards, converting euros to dollars, and translating the app into French, among other tasks.

Kalanick simply directed his team to work harder. "Never ask the question 'Can it be done?'" he was fond of saying at the time, recalls one employee. "Only question how it can be done."

Kalanick left for LeWeb but stayed in touch from his hotel room over Skype video chat, his disembodied head still a loud, demanding presence in the office. Everyone was working around the clock, on little sleep and ebbing patience. "Someone turn Travis off!" yelled the new chief of product, a former Google manager named Mina Radhakrishnan, when Kalanick berated them for not having the service ready in Paris on time.

Conrad Whelan, the company's first engineer, recalls spending every day in the office, from 7:30 a.m. to midnight, including weekends, for three weeks straight before the Paris launch. "This is the biggest thing I will say about Travis," he told me years later. "He came to us and said, 'Look, we are internationalizing and launching in Paris,' and every single engineer was saying, 'That is not possible, there is so much work, we will never be able to do it.' But we got it done. It wasn't perfect. But that was one of those moments where I was like, 'This Travis guy, he is really showing us what is possible.'"

Kalanick introduced the service onstage at LeWeb as planned. Uber's investors were left in awe and with a little bit of trepidation. Opening in Paris then "made no sense, none," says Chris Sacca, the cowboy-shirt-wearing angel investor, at the time one of Kalanick's closest advisers. "We hadn't even done Los Angeles or Houston or these huge black-car markets. It was pure gumption—a moment that demonstrates the difference between investors and one of the greatest entrepreneurs in the world. We could see all the reasons not to make that move and Travis just knew it was going to work anyway."

Throughout 2011, Kalanick ruminated over the lessons of the past year's experiment with price hikes. Uber had doubled rates for riders in San Francisco the previous New Year's Eve in a bid to encourage more drivers to stay on the road during a frenzied night. Enticing additional drivers onto the road with richer paydays while discouraging hails from poorer riders could bring supply and demand into balance during peak times. It could also help solve one of the biggest problems of the taxi industry: the utter unavailability of cabs when they are most needed, such as on boozy weekend nights, during holidays, or when it rains.

"On a New Year's Eve or Halloween or big music festival, demand just goes nuts," Kalanick explained on the podcast *This Week in Startups* that August, before there was even such a thing as surge pricing on the Uber app. "It gets to a situation where people are pushing the button and they have to do it like twenty times to get a ride. So you raise the prices to lasso in demand. It's classic economics."[5]

Not everyone inside the company agreed with the rationale or with the plans that took shape that year to experiment with a broader roll-out of dynamic pricing. Many of Kalanick's employees thought temporarily raising rates might alienate customers and wouldn't necessarily motivate drivers. There was also debate over what to call such proposed rate changes, recalls Ryan Graves. *Dynamic pricing* wasn't quite right, Kalanick argued, because prices would never fall below the baseline rate. The term *surge pricing,* he believed, was more accurate, plus it sounded slightly foreboding, which was precisely the point. "It was supposed to be a little bit scary," says Graves, to encourage some passengers to look for other transportation options.

Uber ran another test of surge pricing over Halloween, again capping the increase at two times the normal rate.[6] The process was manual; the general managers of the six city offices converged in a Skype chat room that night and monitored the fleet of vehicles using the service. If cars got scarce and the managers wanted to raise prices, they requested the higher rate. In San Francisco, Kalanick plugged the new fares into the software.

But to really establish an equilibrium between supply and demand during such frenzied nights, Kalanick reasoned, Uber was going to have to remove the cap altogether and let the almighty hand of the market figure out the price. All internal objections were overruled. The company's overarching goal was to have cars always available, at any time of day or night, and surge pricing could help Uber achieve that.

On New Year's Eve, Kalanick and most of the engineers decamped to Costa Rica for another holiday workation. By then an engineering team led by Kevin Novak, previously a nuclear physics researcher at Michigan State University, had created an algorithm that automatically upped fares in reaction to a scarcity of available cars. From the beach, Kalanick and his colleagues watched as the first experiment in uncapped algorithmic surge pricing played out in real time. It was a disaster. "We knew it was going to be tough, but we didn't know it was going to be that tough," Kalanick says. "I mean, I had seventy-two hours of just..." His voice trails off at the memory.

After midnight, prices spiked seven times the normal rate in New York and San Francisco. Passengers were paying more than a hundred dollars for relatively short rides. Enraged customers flocked to social media to complain. Even though the Uber app had displayed surge multipliers like *1.8x* or *2.5x* to users, customers either didn't see it or couldn't quite comprehend what the number meant. Uber had its first serious public relations crisis on its hands. *While I'm glad I'm home safely, the $107 charge for my @Uber to drive 1.5 miles last night seems insanely excessive,* Tweeted one New Yorker.[7]

Kalanick watched all this unfold from Costa Rica and succumbed to his first instinct, which was to respond combatively and defend his beloved brand. *The price is right there before you request... it's simply an option... those who select it are making a choice on how to spend their money,* he Tweeted to one user.[8] He told someone else who was charged sixty-three dollars for a three-minute ride: *The sticker shock is rough tho our records show u saw the price increase notification 4 separate times b4 requesting ride.*[9]

Not surprisingly, blaming customers didn't help the cause. Stories about the perceived price gouging appeared in the tech blogs and major outlets like the *New York Times* and the *Boston Globe.* Kalanick pointed out that gas prices had historically surged based on supply conditions and that people were just going to have to deal with it and overcome seventy years of conditioning around fixed prices in ground transportation.

Privately, Uber execs knew they hadn't done a good job handling the situation. Allen Penn, a manager in the Chicago office and a friend of Ryan Graves from college, said the company was still learning how the algorithm would impact prices and how customers might react. "Our messaging was not that great," he says. "We didn't do a great job of communicating to people what it would cost." Even Kalanick, somewhat contrite after the media maelstrom, told me a few months later that the details of communicating price increases, even font sizes and wording, mattered a lot. "We tried to unwind decades of fixed pricing in personal transportation in one night," he said. "There was a little angst around it."

At least one investor, Chris Sacca, was fuming over the way Kalanick had handled the media deluge and compared the response to Mark Zuckerberg's retort to Facebook users after they protested the initial rollout of the newsfeed in 2006. (*Calm down,* Zuckerberg had started his blog post.) "You can't say out loud, 'Fucking get over it,' you have to be like, 'We are working on it, good feedback, we'll improve the app,'" Sacca says.

At the time, Kalanick seemed convinced surge pricing was a tool only for special occasions. "I don't think that the constantly changing car price is necessarily where we want to go," he told the *New York Times*. "But on Halloween and New Year's, it's here to stay."[10]

Then one of his own colleagues helped to change his mind. Michael Pao was a recent Harvard Business School graduate who had overcome Kalanick's allergy to hiring MBAs and caught on with Ryan Graves's operations team. He spent a few weeks working in Chicago then moved to Boston in October, where he met Austin Geidt and tried to hire a local team. When they couldn't find a suitable general manager, Pao took the job himself.

Pao had lived in Boston for six years and was familiar with the inconvenience of the city's weekend rhythms. Most bars in Boston closed at 1:00 a.m. On Friday and Saturday nights, their drunken clientele staggered onto the street in unison. Cabdrivers wanted nothing to do with this ritual and would promptly head home at that hour to keep their lives free of drama and their backseats free of vomit.

Uber drivers, naturally, were exhibiting the same behavior. After thinking it over and fretting he would never be able to grow Uber's business in Boston if he couldn't solve the challenge of closing time, Pao started running experiments. For a week he held fares steady for passengers but increased payments to drivers at night. In response, more drivers held their noses and stuck around for closing hours. It turned out drivers were in fact highly elastic and motivated by fare increases. Pao then spent a second week confirming the thesis by breaking Boston's Uber drivers into two groups. Some saw surging rates at night while others did not. Again, drivers with higher fares during peak times stayed in the system longer and completed more trips.

Pao now had something that the previous unfocused tests of surge pricing hadn't yielded—conclusive math. He presented his findings to Kalanick, showing that by offering more money to drivers at certain times, he could increase the supply of cars on the road by 70 to 80 percent and eliminate two-thirds of unfulfilled requests.[11] Kalanick was convinced, and surge pricing became orthodoxy inside Uber, despite the negative response to the New Year's trial. After that, regardless of the avalanche of criticism from the media, the hostility from regulators, and the unpopularity of surge pricing among riders, Kalanick never equivocated on it again. The math was on his side.

"Our principles are clear," he told me in 2012. "Number one, Uber is always a reliable ride. Always. You can't say that for some of the alternative transportation systems in the city. In fact, probably none.

"Number two, we only implement dynamic pricing, or surge pricing, if it will increase the number of rides that happen. When prices go up, more drivers come out. When more drivers come out, more rides happen. That means less people are stranded and more people have an option."

That, of course, was only part of the story. Uber was addressing the chronic shortage of cars during spikes in demands by tailoring the service to people who could afford to pay extra. There was a kind of cruel economics at play, and riders would continue to have visceral resistance to the idea that the same ride could cost more at different times. Observers would connect the tactic with Kalanick's Twitter avatar—at that point, the cover of one of Ayn Rand's manifestos, *The Fountainhead.* "It's less of a political statement," Kalanick told the *Washington Post* reporter who asked about it in 2012. "It's just personally one of my favorite books. I'm a fan of architecture."

But Kalanick's stubborn defense of surge pricing impressed at least one observer. "Travis is a real entrepreneur," Amazon CEO Jeff Bezos told board member Bill Gurley after a surge fracas, according to Gurley. "Most CEOs would have caved."

In the fall of 2011, Travis Kalanick once again set out to raise capital. The rancor surrounding surge pricing would be nothing compared to the animosity that was about to be unleashed behind the scenes.

Though still small, Uber was showing flashes of promise. In September, it generated $9 million in fares and kept $1.8 million in commissions, according to data shared with investors at the time. It had nine thousand customers using the app, with 80 percent in San Francisco, though other cities were growing quickly. A seductive salesman in investor meetings, Kalanick laid out an enticing vision: Uber could become a global brand, like FedEx, with the potential to introduce new kinds of car services at lower prices. "All the numbers on those charts blew me away," says Gary Cohn, president of Goldman Sachs. He had met Kalanick in the Market Street office and later persuaded his company to pitch in five million dollars, the initial bond in what would become a close relationship between the investment bank and the startup.

Not everyone devoured the pitch. Venture capital firms like Yuri Milner's DST took a look but passed, reasoning that Kalanick was nothing like the introverted CEOs of Facebook and Google. A few other firms expressed interest, but Kalanick's clear favorite was the newest sugar daddy on Menlo Park's Sand Hill Road: Andreessen Horowitz, the two-year-old firm that a few months before had led the Series B round in Airbnb, making the home-sharing startup a unicorn.

The attraction for Kalanick was the same as it had been for Brian Chesky. The firm was led by entrepreneurs Marc Andreessen and Ben Horowitz and known for offering favorable terms at muscular valuations. Like Chesky, Kalanick wanted to enlist the services of Andreessen's newest partner, Jeff Jordan, an expert in the peculiar dynamics of online marketplaces.

At first, Andreessen Horowitz was the most aggressive pursuer, offering to value the startup at more than $300 million. But then Marc Andreessen, the Netscape co-founder, changed his mind and, at dinner with Kalanick, told him that Uber's financials didn't yet support such a rich valuation and changed it to $220 million, according to a report in *Vanity Fair*.[12]

Though disappointed, Kalanick tentatively agreed to the new terms but then felt further blindsided when he saw the fine print. Anticipating an influx of new hires, Andreessen Horowitz wanted to create a large option pool at Uber (the shares to be doled out to new executives and employees). That meant the shares of older investors and employees would be further diluted. It was a critical mistake. Kalanick now felt he was getting shortchanged. Fortunately, he had a plan B.

The Iranian-born Shervin Pishevar, a partner at one of Silicon Valley's oldest venture capital firms, Menlo Ventures, had also been competing for the deal. Pishevar was a bearded bear of a man who doled out hugs easily and was prone to frequent bouts of sentimentality. He had a mixed track record as a startup founder himself, but as a venture capitalist, Pishevar represented an emerging class of investor in Silicon Valley. Instead of hard-won experience and business wisdom, he had social connections and charisma. He was a cheerleader as much as a thought leader, able to align himself with fashionable new ideas and more than willing to demonstrate his support publicly, doing everything from sending Tweets to shaving the logos of his portfolio companies into his hair (something he did twice). Pishevar, a consummate networker on both coasts, also offered something that Marc Andreessen could not: access to celebrities and politicians. That would come in handy.

Kalanick was fond of Pishevar and had let him down gently when he decided to go with Andreessen Horowitz. Pishevar amiably told Kalanick to call him if anything went wrong during the closing of the round. When it did, Kalanick picked up the phone from a technology conference in Dublin and asked Pishevar if Menlo Ventures was still interested. Pishevar was speaking at an event in Tunisia and nursing a bad back. Nonetheless, he promptly got on a plane.

In Dublin, Kalanick and Pishevar walked the cobblestone streets, got beers, and talked about Uber's future. Sensing a huge opportunity, Pishevar offered a $25 million investment at a valuation of $290 million and didn't even ask for a seat on the Uber board, which meant Kalanick could postpone the inevitable shift of influence on the board to his investors. Pishevar says he was impressed by Kalanick's insane devotion to the company and by the addictiveness of the service. Each Uber user at the time took three and a half rides a month and showed the app to seven friends, Pishevar recalls. "With those numbers I estimated they would hit a hundred million dollars gross revenue in a year," he says. "They did it within six months."

Kalanick now had a big decision to make and sought advice. Michael Robertson, a San Diego music entrepreneur who knew Kalanick from his Scour days, recalls getting a call from Kalanick that week. The Uber CEO explained that he was being offered a good deal from a practically unknown investor (Pishevar) and a less valuable deal from considerably more famous one (Andreessen). What should he do? "You don't need validation from a venture capitalist. You are past that," Robertson told him. "Now it's about getting the cheapest capital you can. Capital is power. The more capital you have, the more options you have."

Kalanick took that advice to heart and signed a term sheet with Pishevar outside his hotel room at the Shelbourne in Dublin. On Friday, October 28, he e-mailed Garrett Camp, Uber's other board members, and lawyers at the law firm Fenwick and West and informed them of the deal. "If we haven't talked in the last 24 hours you may be wondering what happened to Andreessen Horowitz," he wrote in the e-mail. "Well, they tried to surprise us with a large option pool on an already low-priced round ($220mm pre) that when you did the math, the numbers didn't work. So here we are. The next phase of Uber begins."

The investment shuffle would have broad ramifications for Uber and, later, for one of its emerging competitors. As Andreessen Horowitz realized the magnitude of its mistake, it would lead one of the earliest fund-raising rounds in Lyft. Uber's deal with Pishevar would also lead, indirectly, to the collapse of one of Kalanick's closest friendships.

Over the next few months, Pishevar worked his vaunted network and enlisted the support of big-name Hollywood stars and Silicon Valley celebrities. Uber's new investors included the actors Sophia Bush, Olivia Munn, Edward Norton, Ashton Kutcher, and Jared Leto; the performers Jay Z, Jay Brown, and Britney Spears, along with her former manager Adam Leber; the talent agency William Morris; and the music manager Troy Carter. From high tech, he helped bring in Jeff Bezos and Google CEO Eric Schmidt. All these luminaries put in between $50,000 and $350,000 each. By 2016, their stakes had grown twenty times.

Someone else ponied up too: Brian Chesky. The Airbnb founder says Kalanick himself invited him to invest in the round. "I knew the company would be massive. I didn't know how big it would be," he says.

Some of Uber's earlier investors were suspicious of this rolling, seemingly endless financing and particularly concerned by the fact that the celebrities were given the same deal terms long after the round had closed. By then Uber's growth in new cities was accelerating. Uber was going to be huge, and that ancient mortal sin greed was beginning to rear its head.

Chris Sacca recognized the sheer size of the possible opportunity sooner than most. Uber's earliest angel backer had built an impressive investment record supporting early startups and then selectively doubling down, buying additional stock from fellow investors who wanted to cash out or had less of an appetite for the continued risk. He had made a massive haul doing this with Twitter, in part by staying close to Twitter co-founder Ev Williams.

Now he pursued the same strategy with Uber. Kalanick appeared amenable, at first, but then seemed to change his mind. Sacca attempted to reacquire the shares Uber had sold to Universal Music in 2010 for the rights to the name Uber.com but Kalanick had beaten him to it and already bought them back for the company. Sacca reached an agreement with several early investors to buy a portion of their stock but he needed approval from Uber to complete the transaction. Kalanick refused to provide it, worrying that it would change the market price at which the company could grant stock compensation to new employees. Kalanick also believed Sacca was trying to *sell* Uber stock. Sacca vehemently denied this.

The two had been close friends for years, spending hours brainstorming in the Jam Pad, soaking in Sacca's hot tub in San Francisco, and vacationing in Sacca's house in Tahoe. Sacca had brought Kalanick and Garrett Camp to Washington, DC, for the Obama inauguration. But now there was only Uber. Kalanick was betrothed to the company and its magnificent potential. Sacca, it seemed to Kalanick, was out only for himself.

Tensions between the two mounted during 2011 over the issue of secondary stock and then came to an explosive head after the conclusion of the financing round with Shervin Pishevar and his celebrity friends. Kalanick needed Sacca to sign a set of closing documents. Sacca says that at the time he was spending sleepless nights with his newborn baby and that he signed them without fully reading them. The documents, it turned out, contained an agreement to take away some board rights from First Round Capital.

When Sacca found out what he had signed, he says he was furious. Josh Kopelman of First Round, Rob Hayes's partner, had helped him get his start as an angel investor. Now he felt like he had undermined him. Early investors often voluntarily give up rights such as board seats and the ability to invest in future rounds, but they are loath to have them stripped away. "Dude, I have to live in this industry!" Sacca complained to Kalanick.

Soon after, Kalanick stayed overnight at Sacca's house in Santa Monica. Sacca brought it up again as they were talking in the kitchen and Kalanick replied icily, "You should learn to read documents before you sign them." Sacca and his wife kicked Kalanick out of their house.

Sacca continued to attend board meetings as an observer but the end was near. There are slightly varying accounts of how things unfolded but they all start in September 2012, with a conversation between Sacca and Pishevar. According to Sacca, they were talking about how to support Kalanick's growth as a CEO when Sacca mused that a different kind of investor might have sued over an issue like the document signatures.

According to Pishevar, Sacca was more direct. He said he felt pressured to sign the documents and that situations like that might leave him no choice but to sue Uber.

There is complete agreement about what happened next. Pishevar promptly called Kalanick and reported the real or imagined threat of a lawsuit. Kalanick then called Sacca. "He was yelling and screaming," Sacca recalls. " 'You are going to fucking sue me! Fuck you!' "

A few weeks later Sacca was preparing to attend a scheduled Uber board meeting. Kalanick bluntly told him he wasn't welcome. Sacca said he was going to come anyway and wanted to "talk it out." Kalanick informed that if he did, security would escort him away. The law firm Fenwick and West then sent Sacca a letter stating that he could no longer attend Uber board meetings as an observer and was not entitled to any private information about the company.

Over the next few years, Sacca sent repeated e-mails of apology and made numerous attempts at reconciliation. In a profile in *Forbes* magazine, he even framed the rift, only half truthfully, solely as a result of his efforts to buy more Uber stock.[13] But, as of the publication of this book, Kalanick and Sacca had yet to settle their differences.

Austin Geidt finally returned to San Francisco for a few weeks in early 2012 after Josh Mohrer, a former marketing director at an online wine seller, signed on as the new GM for New York City, replacing Matthew Kochman. Thinking she was going to stay at home for a while, she got a dog—a mutt named Dewey. Then she was suddenly on the road again, opening Los Angeles and Philadelphia, toting the dog everywhere, looking for a hotel that would take him and trying, unsuccessfully, to crate-train him. "It wasn't the ideal time in life to get a dog," she says.

Big sprawling L.A. was the perfect place for Uber to run its playbook. The company had a launch party at SmogShoppe, a former auto shop that had been converted into a restaurant, with celebrity guests that included Olivia Munn, Ashton Kutcher,

NFLer Reggie Bush, and model Amber Arbucci. The actor Edward Norton, a friend of Pishevar's, took one of the first rides, which the company advertised in a blog post. Before long, the city was buzzing about the hottest new startup and its celebrity coterie.

Employing the SoHo strategy, Uber launched in Hollywood and Santa Monica first, with guaranteed minimum daily payments to drivers. When the service started to gain momentum there, Uber switched to a straight 20 percent commission on all rides in those neighborhoods and moved daily minimum guarantees to other parts of the city. In this way, Uber used its bank account to fuel its growth. "If we had tried to launch L.A. all in a day we would have failed miserably," Geidt says.

Uber was now spreading with urgency across North America. By early 2012 it was in a dozen cities and had fifty employees, about half in the field. With venture capital in the bank and a host of potential rivals suddenly looming on his radar, Kalanick was ready to hit the accelerator. "I was waiting for Travis to say slow down," says Geidt, "but that never happened."

CHAPTER 8

TRAVIS'S LAW

The Rise of Ridesharing

I'm an idealist, it's always been a problem of mine and I do apologize in advance.

—Travis Kalanick, open letter to Washington, DC, city council members[1]

Until this point in its brief but eventful history, Uber had moved with relative caution into new cities. Though Travis Kalanick and his colleagues had come to distrust taxi ordinances as schemes designed to protect incumbents and their shoddy levels of service from new competition, they examined local laws closely and were flexible when required. Uber was by and large a law-abider, not a law-bender. Over the next two years, and for surprising reasons, that changed.

In 2012, the company would come face to face with steely regulators, an international rival with aggressive expansion plans, and, unlikeliest of all, possible disruption from two other Silicon Valley upstarts that were ready to discard taxi laws altogether. These events would bring out Kalanick's remarkable adaptiveness as well as his fierce competitive streak, with heady ramifications for the company, American cities—even the world.

It started with a Tweet.

On January 11, 2012, at 10:35 a.m., a short, cryptic message from a rider-advocacy group called DC Taxi Watch quoted the top taxi official in the U.S. capital. *Chairman Linton: @uber DC is operating illegally,* it read.

The Tweet was sent from inside the drab, postwar DC Taxicab Commission headquarters in Anacostia. The city's taxi drivers had packed a normally sleepy hearing to make their voices heard. Uber's town-car drivers, they argued, had been illegally operating for the past two months.

Ron Linton was inclined to agree. Appointed only six months before by Mayor Vincent C. Gray to head the taxicab commission, Linton was in his early eighties. He was an avuncular policy planner and longtime reserve officer in the city police department who wore a stern disposition and an obvious toupee. Like his counterpart Christiane Hayashi in San Francisco, he fashioned himself an agent of change and was determined to modernize the capital's pitifully antiquated taxis, which ignored minority neighborhoods and didn't accept credit cards. They didn't even have dome lights or a uniform color to distinguish them from other cars. But Linton was hell-bent on doing it from inside the industry and preserving the jobs of the region's eighty-five hundred licensed drivers. Uber is "operating illegally, and we plan to take steps against them," Linton assured the boisterous drivers at the meeting.[2]

Uber's DC general manager, Rachel Holt, was just settling into her new office when she saw the Tweet from the hearing. As in the other cities Uber had entered, the maze of local taxi regulations didn't seem to explicitly prohibit the company's service. In DC, yellow cabs had to use taximeters to calculate fares, while limos could only charge a prearranged fare. But there was a third classification in the bylaws, under section 1299.1 in the District of Columbia Municipal Regulations, which seemingly contradicted the other two rules by stipulating that sedans carrying six passengers or fewer could charge on the basis of time and mileage.[3] Uber's approach clearly qualified.

Holt was a former Bain consultant and marketing manager at consumer-goods maker Clorox, in Oakland, whose fiancé worked in DC. When she started looking for a job in the capital, she had one important condition: "The only thing I knew I didn't want to do was politics," she says. A friend showed her a job posting at Uber, which was looking for someone to lead the rollout in Washington. After meeting Graves and Kalanick, she says she became excited about the autonomy that came with being "CEO of a city" and working for a young, promising startup. She spent her first month in San Francisco, then logged a month at the office in New York, learning the ropes and helping Graves and Geidt to reframe the New York strategy. After that, she moved to DC. Uber started facilitating rides there in November 2011 and officially launched in December. It didn't take long to get the region's cosseted taxi drivers, unaccustomed to new competition, steaming mad.

After she saw the Tweet from the taxicab commission hearing, Holt e-mailed Linton's office asking for a clarification. She was told she would hear back within forty-eight hours. That was a Wednesday. Ron Linton was true to his word. On Friday, his office tipped local press to assemble outside the Mayflower Hotel on Connecticut Avenue. The chairman then ordered an Uber town car from the Cleveland Park neighborhood and took it to the hotel, where he was met at the circular driveway by five hack inspectors from the DC Taxicab Commission.

Surrounded by three reporters, the officers slapped the stunned driver with $1,650 in fines for driving an unlicensed vehicle in the District and not having proof of insurance on hand, among other infractions. Then they impounded his car for the Martin Luther King Day long weekend. Standing in front of the press, Linton slammed Uber for unleashing regulatory havoc in the city. "What they're trying to do is be both a taxi and a limousine," he said. "Under the way the law is written, it just can't be done."[4]

Holt, who had arrived three minutes late to the scene after being alerted by the driver that trouble was afoot, was perplexed. According to the actual citations, Linton was going after the driver himself, a Virginia resident, not Uber, and he was doing it based on one of the city's more arcane and senseless rules—that limo drivers must present a fare to the passenger in advance, rather than using a meter that measures time and distance. The fines did not affect whether Uber could continue to operate in the city and seemed mostly aimed at intimidating drivers and keeping them from signing up with Uber in the first place.

The dispute shifted to the internet. Ryan Graves blogged about the incident on the Uber website, noting the company would be covering the driver's fines and compensating him for a weekend of missed work. "We are surprised that a public official is making statements about Uber violating the law without sending some kind of notice stating specifics," he wrote.[5] He also invited users to Tweet their support and call or e-mail the DC Taxicab Commission directly. It was the first step down a path that would become increasingly important that year: mobilizing Uber's customers to fight on its behalf.

For his part, Linton, who passed away in 2015, suggested that he was protecting incumbent taxi firms as much as enforcing the law. "I'm getting tremendous pressure from cab companies [over] the way Uber is functioning," he said to *DCist,* a local blog, a few days later.[6] "Nobody loves a regulator. We got rules, we got regulations, we got laws." Perhaps aware of the regulatory swamp that he had wandered into, he then referred the matter to the city's attorney general, Irvin B. Nathan, and asked him to evaluate Uber's legal status. That spring Holt met with Nathan and his staff, and officials speculated that the entirety of Section 1299.1, the provision that seemingly gave Uber protection, was nothing more than a

A young Brian Chesky with his parents, Bob and Deb, in Niskayuna, New York. (Courtesy of Airbnb)

A teenage Nathan Blecharczyk at home in Boston, already running a successful internet business. (Courtesy of Airbnb)

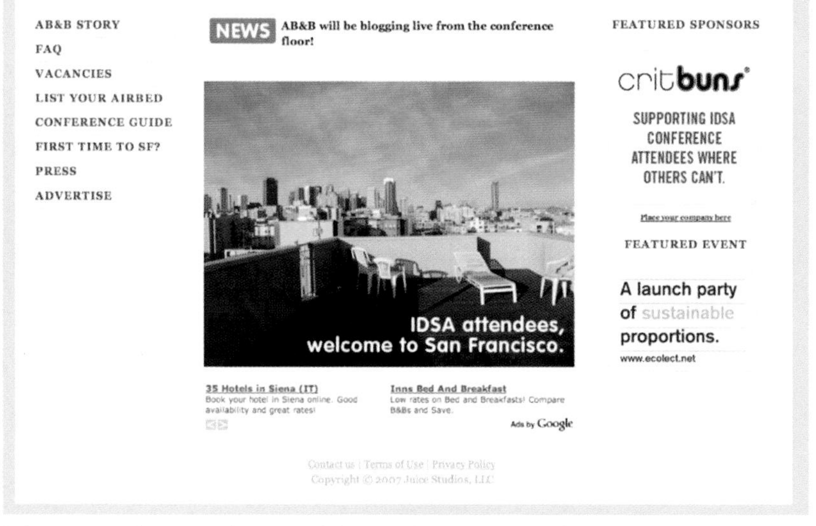

The original AirBed & Breakfast website in October, 2007. (Courtesy of Airbnb)

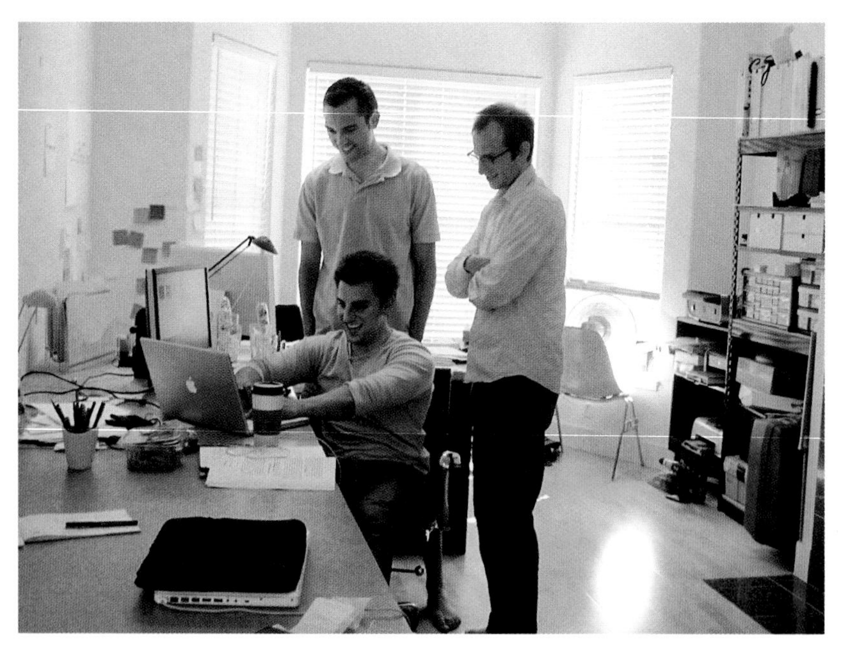

The Airbnb founders (from left: Nathan Blecharczyk, Brian Chesky (seated), and Joe Gebbia in their Rausch Street Apartment. (Courtesy of Airbnb)

Airbnb co-founder Joe Gebbia with his seat cushion, CritBuns, "the ultimate sitting tool." (Courtesy of Airbnb)

The Airbnb founders outside the Y Combinator startup incubator in Mountain View, Calif. (Courtesy of Airbnb)

The Airbnb co-founders (From left: Blecharczyk, Chesky, and Gebbia) in their first offices in 2010. (Courtesy of Airbnb)

A young Travis Kalanick (right) with his father, Don, mother, Bonnie, and brother, Cory. (Courtesy of Uber)

A photo of Travis Kalanick performing the long jump from his 1994 Granada Hills High School Yearbook. (Courtesy of the author)

Travis Kalanick from his 1994 Granada Hills High School Yearbook. (Courtesy of the author)

An early screenshot in 2010 of the UberCab website. (Courtesy of the author)

The early Uber crew: (left to right) Curtis Chambers, Travis Kalanick, Stefan Schmeisser, Conrad Whelan, Jordan Bonnet, Austin Geidt, Ryan Graves, and Ryan McKillen. (Courtesy of Uber)

Early Uber executives Ryan Graves and Austin Geidt pondering a move. (Courtesy of Uber)

Opponents of Airbnb rally before a hearing on short-term rentals at New York City Hall on January 20, 2015. (Shannon Stapleton/REUTERS)

Opponents of Airbnb rally before a hearing on short-term rentals at New York City Hall on January 20, 2015. (Shannon Stapleton/REUTERS)

Airbnb co-founders on stage at a company event in February 2015. (Courtesy of Airbnb)

Allison Chesky, Brian Chesky, and their parents, Bob and Deb, at the Airbnb Open in Paris in November 2016. (Aaron Ke)

Taxi drivers protest Uber by burning tires and blocking streets in Bordeaux, France, on January 26, 2016. (Georges Gobet/AFP/Getty Images)

Uber drivers and their supporters object to lower fares at an Uber office in New York City on February 1, 2016. (Seth Wenig/AP Photo)

Cab drivers protest Uber in downtown Bogota, Colombia, on March 14, 2016. (Fernando Vergara/AP Photo)

Taxi drivers demonstrate against Uber, in Cali, Colombia, on June 28, 2016. (Luis Robayo/AFP/Getty Images)

Uber co-founders Garrett Camp and Travis Kalanick in front of the Eiffel Tower in Paris. (Courtesy of Uber)

Lyft President John Zimmer wearing a frog costume to promote the ride-sharing service Zimride in 2011. (Courtesy of Lyft)

John Zimmer, co-founder and chief operating officer of Lyft, left, and Logan Green, co-founder and chief executive officer of Lyft, have a laugh at the TechCrunch Disrupt SF 2013 conference in San Francisco on Monday, Sept. 9, 2013. TechCrunch, which runs from Sept. 9 through 11, features leaders from various technology fields and includes a competition for the best new startup company. (David Paul Morris/*Bloomberg*)

Didi Founder and CEO Cheng Wei at Didi Offices in Beijing, China. (Ka Xiaoxi)

Didi CEO Cheng Wei and President Jean Liu at a company party in February, 2016. (Courtesy of Zhang Wei / Didi)

Brian Chesky and Travis Kalanick on stage at TechCrunch Disrupt in May 2011 with moderator Erick Schonfeld. (Joe Corrigan/Getty Images for AOL)

Brian Chesky and Travis Kalanick appearing together at the TechCrunch Disrupt conference in New York City, May 2011. (Charles Eshelman/Getty Images for AOL)

typographical error. Uber could keep operating for the time being, but its battle in DC had only just begun.

On Travis Kalanick's trips to Paris and other European cities, one emerging rival in particular was catching his eye. Hailo, a startup operating out of the lower deck of a retired World War II merchant ship docked on the Thames, was supplying a smartphone app to drivers of London's iconic black taxicabs, the equivalent to yellow taxis in the United States.

Hailo was started by Jay Bregman, an American with a master's degree in media and communications from the London School of Economics. In 2003 Bregman had begun outfitting bike messengers with GPS devices to make their routes more efficient. His company, eCourier, was ahead of its time and was decimated by the financial crisis in 2009. Surveying the carnage after eCourier's assets were sold off to a larger competitor and watching Uber's emergence in the United States, Bregman saw an opportunity to use the iPhone to aid London's taxi drivers, who were facing an onslaught of competition from an expanding type of for-hire vehicles, called minicabs, which had to be prebooked on the phone or at a minicab-fleet office. "The insight was to bring these guys into the modern age and give them tools to galvanize them and help them win back work," says Ron Zeghibe, an investment banker who helped Bregman sell eCourier and became Hailo's co-founder and chairman.

Bregman recruited three London cabbies for Hailo's founding team and started pitching the service to the city's grizzled drivers. Because black cabs were already relatively expensive compared to minicabs, the company didn't charge riders anything to use the Hailo app. Instead, it encouraged them to tip and took a 10 percent commission from the fare. Drivers complained, at first, but then the app started delivering new passengers that they wouldn't have found by cruising the streets and looking for people waving their arms in the air. By early 2012, Hailo had been downloaded two hundred thousand times and was being used by two thousand drivers.[7] It was the first of several formidable international rivals that would soon loom large on Uber's radar.

Then Jay Bregman made his first critical mistake. Hailo raised $17 million from the venture capital firms Accel Partners and Atomico and on March 29, 2012, in a poorly calculated display of bravado, it announced plans via the tech-news site *TechCrunch* to expand to the licensed taxi fleets of Chicago, Boston, Washington, and New York—all Uber markets. "Already, Hailo has hired a Chicago general manager and is looking to quickly expand in the coming months," read the last sentence of the *TechCrunch* article.[8]

The news of Hailo's expansion resonated around the world. It was read as far away as China, where, as we shall see later in this account, entrepreneurs and venture capitalists suddenly recognized that linking taxis and limos with riders via mobile apps was an idea powerful enough to cross continents. The problem was that Hailo's introduction into its target cities was actually months away. And inside Uber, where execs read the story closely, the line about Chicago in particular triggered alarm bells. Allen Penn, a manager in Uber's Chicago office, went into war mode, assembling his colleagues on a conference call that very night to discuss a response. The obvious solution was to try to beat Hailo to it—to bring Chicago's yellow-cab fleet onto the Uber app.

That was a serious move, with ramifications for not only the way Uber worked but also the way the company presented itself to the world. Up to that moment, calling an Uber was meant to be classy, upscale, and pricey. An Uber ride at that point was 50 percent more expensive than a ride in a regular taxi. In the estimation of founder Garrett Camp, the name stood for something meaningful—stepping out of a black BMW to meet your friends outside a nightclub was so *uber*. Was it also uber to stand on the corner of Michigan Avenue and Wacker Drive and order a checkered yellow with a funny smell in the backseat?

Over the next few days, there was heated debate inside the company about putting traditional taxis on the system. Uber would have to accommodate their metered fares and stringent licensing requirements and cede most of the commission to the driver to replace the tip and service fees, cutting deeply into its standard 20 percent margin. Many Uber employees and execs opposed the move. "We did this high-end thing, this 'everyone's private driver' experience," says early engineer Ryan McKillen. "We wanted it to stay high end and make it amazing. Doing taxi just felt like so the opposite."

Kalanick eventually put an abrupt end to the discussion with a pointed insight about why Uber was successful in the first place. "I'm going to literally flip this table if anyone says one more time they are worried about destroying the brand," he said in a meeting, according to Penn. "The luxury of Uber is about time and convenience. It's not about the car."

With urging from board member Bill Gurley, who was wary of anyone undercutting Uber on price, Kalanick reached an important conclusion. Uber didn't necessarily have to be a high-end brand. It could compete against all forms of alternative transportation by presenting the most efficient, most luxurious option at any price.

A week after the *TechCrunch* article, Allen Penn was visiting family in Kentucky when Kalanick called and asked him if he could launch a new service, dubbed Uber Taxi, in a week. It would take three. In San Francisco, engineers retrofitted code from an Uber marketing stunt at the recent South by Southwest Conference, which had allowed attendees to order barbecue and pedicabs, and created a new feature to give riders in Chicago a choice between black cars and taxis. On the Chicago streets, Penn and his team started taking taxis, inviting cabbies to the Uber offices, and showing off the app.

Uber rolled out its taxi service on April 18, 2012. Because Kalanick was still nervous about the reception, he framed Uber Taxi as coming from a wing of the still-tiny startup, an entirely fictitious department that he dubbed Uber Garage.[9] "Google has Google X, and we have the Uber Garage," Kalanick told me that year. "If we have an idea we don't like, we put it in the parking lot."

Uber beat Hailo to Chicago's taxi fleet by a mile. The London startup wouldn't open for business there for five more months.

But that wasn't the only reason Uber drove circles around its first major international competitor. There was also a sharp difference in strategy, which was on stark display a few weeks later when Kalanick and Jay Bregman shared a stage at the LeWeb conference in London's Westminster Central Hall, in a panel billed by organizers as a taxi-app smackdown between the CEOs and their backers. Bregman brought with him to the stage one of his investors, Adam Valkin from Accel, while Kalanick brought Shervin Pishevar, who started the session by having the Uber logo shaved into the hair on the back of his head.

Both entrepreneurs politely defined their differences. Hailo was linking the existing supply of licensed taxis, trying to make cabbies more productive by filling in the margins of their day. Uber (aside from the Uber Taxi experiment) was trying to build an entirely new network of professional drivers with luxury vehicles. Hailo drivers could swerve to the side of the road to pick up passengers hailing them from the street; Uber drivers, legally, couldn't.

Then they started throwing rhetorical punches. "We haven't built our product around a market, we've built an experience around a customer desire," Kalanick said, unaware that he hadn't taken the tag off the sleeve of his new sport coat. "That's probably the fundamental difference."

Bregman touted the benefits of using licensed cabs and pointed out the obstacles that Uber was still encountering in New York, where wait times could still be over five minutes. By giving the apps to cabbies, Hailo could "oversupply first and provide great service from the get-go and then it only gets better as our number of cabs increase and the number of customers increase."

Kalanick noted calmly that in places like New York, the number of cabbies on the road at any time was restricted by many factors, such as the limited supply of medallions, shift changes, and spikes in demand. "You need to have flexible supply and sometimes that's when a new network can really come into play," he said.

By the end of the panel discussion, there was little resolution about which approach was better. But a few years later, I caught up with Bregman at a café in the West Village. He had left Hailo after bringing it to North America and getting thoroughly trounced by Uber.

"We thought we would be getting displacement business—people who would stop hailing off the street and start using the app," he told me. "What actually happened was that people stopped driving their cars and renting cars during travel and instead started using ride-hailing apps." As demand boomed, yellow-taxi apps couldn't keep up supply—just as Kalanick had predicted.

Hailo later tried to pivot in London, adding a minicab option to its app, just as Uber had introduced a taxi option in Chicago. But it didn't work. Hackney drivers felt betrayed and stormed the Hailo offices in protest, attacking Hailo's three cabdriver co-founders on social media and denouncing them as traitors.[10] Hailo then had to retreat from the minicab service as well. "The problem," Bregman told me wistfully, "is that you really have to pick sides."

That summer, Uber considered closely the lessons of its Uber Taxi experiment. In Chicago it had put a cheaper ride next to a pricier one, and, according to internal data, both businesses flourished. Riders, not surprisingly, responded favorably to cheaper rides.

So if Uber's brand was versatile enough to accommodate the germ-infested backseats of yellow cabs, what else could it include? Kalanick and his colleagues produced two answers. One was a fleet of luxury SUVs, to be used by larger parties, that would cost more than traditional Uber black cars. The second was a fleet of four-door hybrids, an option that would cost the rider less than the original Uber offering. The name of the service, UberX, was simply the best the company could come up with. "It was a placeholder. We called it UberX because we couldn't figure out a name for it," says Uber's product chief at the time, Mina Radhakrishnan, who adds that Uber Green and Uber Eco were briefly considered and rejected.

Now, an important clarification: Unlike Lyft and Sidecar, the so-called ridesharing companies that were at that very moment making their debuts in San Francisco, the original UberX accommodated only professional drivers who held taxi licenses. Kalanick envisioned a fleet of black Toyota Priuses to be driven by the same types of licensed chauffeurs who were behind the wheels of other Uber vehicles.

Uber introduced these options on July 4, 2012, with a blog post that promised "Choice is a beautiful thing."[11] Kalanick told the *New York Times* that day, "This is the first big step Uber is taking to go to the masses."[12] The SUVs and hybrids were set for San Francisco, New York, and, soon, Chicago—and Washington, DC.

Uber's business was now growing in DC by 30 to 40 percent each month, surprising even its managers.[13] When Rachel Holt started in Washington, Graves had asked her to get the company's business in the city to $7 million in gross bookings by the end of the year. She hit that goal by April. "I thought, *Gee, this is going well*," she says. But the celebration was short-lived. Uber's growing popularity and plans to roll out UberX were about to provoke another five months of bare-knuckle political brawls.

After ineffectual discussions between Uber execs and Washington's attorney general, the issue of Uber's regulatory status fell into the lap of the DC city councilwoman Mary Cheh, the chairperson of the Committee on Transportation and the Environment. Cheh, sixty-two, was a graduate of Harvard Law and a Democrat who had struggled for years to drag anachronistic DC cabs into the modern age. "Even while Uber was coming around, I was in process of trying to reform the taxicab industry which was in the twentieth, maybe the nineteenth, century," she says. She was also a pragmatist who sought peaceful compromise between many of the powerful local taxi interests in what was turning into a radioactive topic. That spring she sent a letter to Ron Linton and the DC Taxicab Commission, asking them to stop towing Uber cars, and then started working toward a compromise among all the parties who were increasingly incensed by Uber's success.

What was needed, she reasoned, was an unambiguous clarification of Uber's legal status that cut through the contradicting regulations and allowed it to operate in the city. Cheh spent the week after Memorial Day 2012 negotiating with Rachel Holt and Marcus Reese, a colleague of Uber lobbyist Bradley Tusk, as well as with Claude Bailey, a well-known local lawyer and lobbyist that Uber had hired to represent it in the city. She also met with Jim Graham, a bow-tie-wearing councilman from Ward 1 and the most vocal champion of the city's yellow-taxi fleets and their drivers (and whose chief of staff had pleaded guilty to charges that he accepted an "illegal gratuity" to promote taxicab legislation in 2011.)[14]

The result of those talks, Cheh thought, was an elegant short-term compromise that she called the Uber amendments. The regulations, added to a broader transportation bill, would give Uber legal sanction to operate. But they also added a price floor, which required Uber to charge several times the rate of a taxicab. Claude Bailey, accustomed to such compromises but not, perhaps, to Kalanick's brand of fiery idealism, indicated a willingness to accept the deal. Cheh then set a city council vote for July 10 and promised that the provisions were only temporary and would be revisited in the next year. "I tried to explain to them that the provision was nothing more than a placeholder. What I needed was room to maneuver," she says.

What happened next would mold the political tactics of Uber, and many of the tech startups that sought to emulate it, for years to come.

In San Francisco, Kalanick had never fully agreed to a minimum fare. Now, recognizing the approaching competition from companies like Hailo and realizing that services like UberX would require aggressive price cuts, he decided he wanted to fight—to the consternation of his own lobbyists, who had already agreed to Cheh's deal.

Then Kalanick started hurling rhetorical hand grenades, labeling Cheh's proposal a "price fixing scheme" on Twitter and accusing the councilwoman of "doing everything to protect the taxi industry."[15]

But Uber was going to need more than Tweets to sway the DC city council. First, colleagues remember, Kalanick sought the backing of the DC tech community and tried to enlist the support of the online-deals company Living Social, based in Virginia. When they didn't respond, Kalanick decided to go right to his customer base. He sent an impassioned letter to thousands of Uber users in DC, complaining that the city council would make it impossible for the company to lower fares and ensure reliable service. "The goal [of the Uber amendments] is essentially to protect a taxi industry that has significant experience in influencing local politicians," he wrote, basically accusing Cheh and her colleagues of corruption.[16] Then he supplied the phone numbers, e-mail addresses, and Twitter handles of all twelve members of the DC city council and urged his customers to make their voices heard.

The next day he posted a public letter to the council members, writing ominously, "Why would you so clearly put a special interest ahead of the interests of those who elected you? The nation's eyes are watching to see what DC's elected officials stand for."

Mary Cheh was taken aback by the ferocity of the response. Within twenty-four hours, the council members received fifty thousand e-mails and thirty-seven thousand Tweets with the hashtag #UberDCLove.[17] When they arrived for the last session of the summer on July 10, Cheh's colleagues all turned to her in confusion and fear. The amendments, she told me years later, had been "a toss to Jim Graham and the taxi drivers," and now, with the weight of the internet bearing down on the council members, the amendments clearly weren't worth it.

"I didn't want to lose anybody on the bigger bill for this one provision," she says. The price-floor idea was gone by midmorning and an alternative amendment was proposed allowing Uber to operate legally in DC until the matter was revisited at the next meeting in September.

Cheh later compared Uber's reaction to the stubbornness of the gun lobby, with its unwillingness to yield even an inch. But she hadn't seen anything yet. Until that point, she had dealt with Travis Kalanick only from afar.

After that first skirmish, Kalanick started spending more time in DC. Marcus Reese, the lobbyist, says the Uber CEO was charming and persuasive in one-on-one meetings with council members in the historic Wilson Building on Pennsylvania Avenue. Then, in September, Kalanick was asked to testify at the all-day meeting of Cheh's Committee on the Environment, Public Works, and Transportation. Ron Linton's Taxicab Commission had once again proposed a bevy of new restrictions, including one that would ban limo fleets with fewer than twenty vehicles—another seemingly arbitrary dart aimed at the independent drivers who worked for Uber.

Kalanick got plenty of advice from his lobbyists in advance of his testimony. *Play it straight. Just stick to the talking points and don't engage in a philosophical back-and-forth. The real advocacy takes place in other forums. In public hearings, you should be benign and respectful.*

He started testifying at 1:15 p.m. after a morning that had included appearances by Ron Linton, various drivers, and Jay Bregman, who wore a suit and tie and pointed out that Hailo worked harmoniously with regulators in London and Dublin and planned to do so in DC as well. But Kalanick wasn't in the mood for gentle courtship. Facts and intellectual arguments, not charm, were his weapons, and unlike Bregman, he wasn't ready to kiss any political rings. Wearing a blue blazer with a white shirt, he interrupted Cheh's first question with the words "I would disagree with that characterization." Things went downhill from there.

"You wanted to make sure that there was a minimum fare on our services so that only rich people could use Uber, not people of middle income," Kalanick told her.

Cheh pointed out that the proposed and discarded price floor was meant as a way to ensure a peaceful transition to a more permanent arrangement. "I know that you like to cast this as some sort of fight," she said. "Do you understand that? I'm not in a fight with you."

"When you tell us how to do business, and you tell us we can't charge lower fares, offer a high-quality service at the best possible price, you are fighting with us," Kalanick replied.

"You still want to fight!" Cheh said in exasperation. The conversation turned to surge pricing. "I am curious about whether that is somehow a kind of gouging," she said. "If there's more demand, why should the rider have to pay more money?"

Kalanick launched into an explanation of the economy of Communist Russia and how long lines formed at stores for essentials such as toilet paper. "It's because the price of toilet paper was too low," he said. "There wasn't enough supply. Everybody could afford toilet paper, but they could never get it because there were too many people that wanted it and not enough people willing to supply it. And so that's kind of the situation that gets created when you're not able to change price."

"So they didn't have any toilet paper," Cheh said with mock amazement.

"It was a rough situation," Kalanick replied. "Look, price controls by governments, you know, they don't always go well. In fact, I'd say ninety-nine percent of the documented cases don't go well."

"But what I'm trying to figure out is why you get the advantage," Cheh said, recalling standing in a long hot line in 1968 to view

Robert Kennedy lying in state after his assassination and being horrified as vendors jacked up the prices for water. "I'm not sure I fully agree with you that this is really an economic mechanism that makes everybody happy!"

In San Francisco, Salle Yoo, Uber's relatively new chief counsel, was watching a webcast of the hearing. According to Marcus Reese, at around this point in the testimony, she started texting him, asking him to pull Kalanick from the stand as soon as possible. *He is in the middle of a public hearing,* Reese texted back. *I can't just walk up to him and say you've got to go!*

The pro-taxi councilman Jim Graham, sixty-seven, wearing a taupe suit and a gold bow tie, was sitting to Cheh's right. "I'm trying to make a point," he scolded Kalanick. "And the point that I'm trying to make is that if you remain unregulated, and the taxicabs remain increasingly more regulated, there's a fundamental inequity to that." He urged Kalanick to reconsider a minimum fare. "I don't want this city to be [all] Uber. I really don't. Because there's too much of a history of our taxicab industry."

"If you allow competition, what you'll get is a better taxi industry," Kalanick said.

"You can't have competition where one party is unbridled and able to do whatever they please whenever they want to do it and the other party has their hands and feet tied," Graham said. "That's not competition."

"That means drivers are making a better living and riders are getting a better service," Kalanick said. "And that doesn't sound bad to me."

Graham said that many taxi businesses in the District were small businesses. "This is a good thing. This is something we want to protect and nurture. This is not something we want to destroy for the sake of some kind of consolidation into a big company." Kalanick tried to interrupt him. Graham snapped, "May I be a member of this committee, please? Do you mind?"

Kalanick laughed. "Go ahead."

After Kalanick left the stand, Graham, visibly infuriated, suggested reinstating and even increasing a proposed minimum fare. Janene D. Jackson, a deputy chief of staff to Mayor Vincent Gray, came up to Marcus Reese and Claude Bailey and offered her own memorably harsh review of the testimony. "Never bring that guy back here!" she said, according to Reese. Later, Jackson couldn't recall saying that specifically. "The hearing was probably a bad one because I have no recollection of it, except that he pissed almost everyone off," she told me.

And yet. By December, with Uber spreading like wildfire through the capital, and knowing that its users were ready and willing to defend the service, Cheh and her colleagues saw the writing on the wall. On December 4, the Public Vehicle-for-Hire Innovation Amendment Act defined without ambiguity a new class of sedans that could be dispatched via a smartphone app and could charge by time and distance. It passed the Washington, DC, city council unanimously, garnering even Jim Graham's vote, without debate.[18] "The real issue is how receptive the government is to the progress of the people," Kalanick told me a few years later. "It's not about the city council or the government, it's actually about how the incumbent industry is persuading them, let's say, to do what I would consider the wrong thing." Ultimately, he added, "DC was very receptive. But it took them time to see it and feel it."

Uber had flexed its political muscles for the first time, and won. A new tactic was then added to the playbook: when traditional advocacy failed, Uber could mobilize its user base and direct their passion toward elected officials. Uber was not the first company to employ this tactic, but it quickly became among the best at it. In its ensuing first wave of political battles — in places like Cambridge, Massachusetts, Philadelphia, and Chicago — Uber would enlist the support of its customers and, usually, win.

Kalanick had broken every rule in the advocacy handbook. Nevertheless, Uber's lawyers and lobbyists, who had begged him, unsuccessfully, to seek compromise and testify with humility, began to whisper in reverent tones about a new political dictate that contravened all their old assumptions. Travis's Law. It went something like this:

Our product is so superior to the status quo that if we give people the opportunity to see it or try it, in any place in the world where government has to be at least somewhat responsive to the people, they will demand it and defend its right to exist.

That fall there were many reasons to celebrate. Uber had intercepted Hailo before its incursion into the United States, won in DC, and demonstrated the primacy of Travis's Law. With more than a hundred employees and growing fast, the company moved into fifth-floor offices on 405 Howard Street, in the South of Market District. The offices were modest; there were three conference rooms, there was no kitchen, and the elevators were always packed. Drivers routinely gathered in the lobby, waiting to pick up company-provided iPhones. Engineers worked in the office from midmorning until late at night, relieving stress with occasional games of floor hockey around the clustered cubicles. Amid all that hubbub was Kalanick himself, who couldn't sit still and could typically be found pacing the floor of the office.

The company was now on pace to generate $100 million in revenue over a twelve-month period. To mark the occasion, Kalanick rented a few adjacent houses in Tahoe, a four-hour drive from San Francisco, and the entire company retired there for a week. Ryan McKillen, one of the early engineers, remembers sitting on a porch overlooking the lake with Conrad Whelan. "Years from now people are going to say, 'I was at Uber Tahoe,'" Whelan told him.

McKillen calls it "an overwhelming moment," the idea that the tiny little startup whose employees were once crammed into a conference room was now changing the world. "It was crazy," he says. But the true craziness hadn't even begun.

Back home, even more momentous events were unfolding. It

had started with something in the air—an obvious idea, perhaps, to those who were carefully studying the Uber phenomenon and were able to see its logical conclusion. The idea appealed to both risk takers willing to ignore decades of strict transportation law and idealists who believed that the idea was so powerful and necessary that policy makers would simply have no choice but to bend the laws to accommodate it.

The idea was this: Until that point, Uber allowed only licensed chauffeurs and taxi drivers to use its system. But what if you opened the service to anyone with a car and allowed him to pick up passengers looking for a ride via a smartphone app? You could fill empty seats in cars, reduce the chronic congestion on America's highways, and allow drivers to make money on the side. It could be carpooling on a mass scale—a digital manifestation of organized programs like 511.org in California or the Sluglines in DC, where drivers stopped in designated lots to pick up passengers so they could access the HOV lanes.

A decade earlier, this might have been called something like mobile hitchhiking. But its originators, carefully trying to fit the idea within the legal protections that state law afforded to casual carpooling, devised a more innocuous term: *ridesharing*.

Internet-enabled ridesharing already existed, amorphously and unremarkably, well before it became a massive moneymaking opportunity. Ridesharing was a popular standalone category on Craigslist in many cities and on the labor marketplace TaskRabbit, founded in 2008, where requests for rides to the airport constituted 10 percent of the early traffic, according to its founder Leah Busque.

In 1997, Sunil Paul, a native of India and the founder of an anti-spam company Brightmail, had the intuition that a phone could one day be used to facilitate rides between people traveling in the same direction. He was granted a patent for his System and Method for Determining an Efficient Transportation Route by the U.S. Patent and Trademark Office in 2002.[19] Paul sold Brightmail to the PC security vendor Symantec in 2004, spent a few years as a venture capitalist, and then, inspired by the successes of Uber, co-founded a company in San Francisco called Sidecar.

Sidecar started giving rides in February 2012 via its apps for the iPhone and Android smartphones. Though it went out of business in 2016, outfinanced and outmaneuvered by Uber and Lyft, it can lay claim to being a pioneering ridesharing company.[20] Anyone—not just taxi drivers or licensed chauffeurs, but your uncle Frank in his beat-up 2008 Accord with the bad paint job—could start driving, as long as he or she cleared an online background check, showed a driver's license and proof of insurance, and maintained a favorable rating from passengers. At first, riders using Sidecar were not required to pay a fee but encouraged to make a suggested donation to the driver, with Sidecar taking a 20 percent cut. It was an attempt to align the service with casual carpooling rather than for-hire taxis. "Our vision is that your smartphone becomes as powerful as your car to get around," Paul told me that year.

But in the long run, no company became more associated with the ridesharing idea, or more of a looming threat in the eyes of Kalanick and his colleagues, than the sharing-economy pioneer and near nonstarter known as Zimride. After a four-year slog, Logan Green and John Zimmer's long-distance carpooling service had contracts with dozens of universities and several companies to use customized versions of its website, and they had a bus service that ran between a few major cities. "We had grinded out a multimillion profitable business," says Zimmer.

But it wasn't growing rapidly, and it wasn't satisfying Logan Green's idealistic dream—which had originated on that college trip to Zimbabwe—of filling the seats in the mostly empty cars that clogged the world's highways. Zimride also wasn't Uber, the company that had demonstrated the remarkable power of smartphones to make transportation within cities more efficient and reliable.

In the spring of 2012, with Uber taking off in cities like Chicago and Washington, DC, the Zimride founders and a few employees started brainstorming new products. One idea was a way to share pictures from road trips; another was a way for people to use their phones to share their locations with family and friends. But the third idea, originally dubbed Zimride Instant, caught the imagination of everyone there. Wherever drivers were going, they could use the company's apps to pick up passengers, not just between cities but inside them.

The notion was discussed at a Zimride board meeting at the company's offices at 568 Brannan Street. Board members wanted to know: Was this even legal? Kristin Sverchek, a partner at the law firm Silicon Legal Strategy and Zimride's outside counsel at the time (she would join the company a few months later) could have stopped the whole thing. She didn't, pointing out that taxi regulations had been crafted decades before smartphones and internet ratings systems were invented. "I was personally always of the philosophy that the great companies, the PayPals of the world, don't get scared by regulation," she told me. "I never wanted to be the kind of lawyer that just said no."

Engineers started working on the system, which was renamed Lyft, after a suggestion from a design intern named Harrison Bowden. Lyft would have the same primary ingredients as Sidecar—a suggested but voluntary donation, driver and passenger ratings, the background check—and it followed Sidecar into public tests in San Francisco by three months. It could have easily been dismissed as an uninspired follower but instead was greeted as a novelty, primarily because Zimmer and Green had thought carefully about a new set of rituals needed to turn a ride with a stranger into a comfortable and safe experience.

The Lyft founders devised a sort of mating dance for strangers sharing a car. Riders were told to climb into the front seat, not the back. They were instructed to exchange a fist-bump with the driver. Conversation was encouraged—everyone here was a fellow passenger on a new internet wave that was connecting people and communities through supposedly superior transportation alternatives. "Getting into someone's Honda Accord was not a normal thing to do," Zimmer says. "It's the thing your parents told you never to do. We had to think about the whole experience."

In a grand flourish, Zimmer decided that every driver using Lyft should affix a pink carstache to his or her vehicle's front grille. The carstache, a jumbo furry mustache recently popularized by another San Francisco company as an eccentric car accessory, had been an inside joke at Zimride, the gift that employees gave away at marketing events and hung on their cubicle walls. Zimmer decided it could be a brand icon and would help turn an otherwise intimidating vehicle into a warm and inviting Lyft car. Carstaches could also command attention. To live in San Francisco in the year 2012 was to wonder, with mounting curiosity, why those weird pink carstaches were suddenly everywhere.

The Zimride founders would not, even under protracted torture, admit that the user interface of archrival Uber was an inspiration for their own design (or vice versa, for that matter, though both companies clearly drew heavily from each other's product features and rhetoric). In the Zimride founders' view, Uber and Lyft were entirely different. "We didn't think of them as similar to us," Zimmer told me. "Our vision has always been every car, every driver, and never 'everyone's private driver.' We didn't want to be a better taxi; we wanted to replace car ownership."

But Kalanick saw through that and knew immediately that the services were competitive. Lyft had some good ideas; only after the carstaches appeared around town, for example, did Uber begin furnishing its own drivers with a windshield decal of the Uber logo.

In a historic irony, though, Kalanick fervently believed that services using unlicensed drivers were against the law—and would get shut down. "It would be illegal," he said on the podcast *This Week in Startups* even before Lyft and Sidecar had launched. "Unless the driver had what's called a TCP license in California and was insured."[21]

"You don't want to get into that kind of business?" asked host Jason Calacanis, the Uber angel investor.

"The bottom line is that we try to go into a city and we try to be totally, legitimately legal," Kalanick replied.

Kalanick was in something of a catch-22. If he responded with his own unlicensed ridesharing service and it was declared illegal, the larger and older Uber could face a much heftier penalty. But if he did nothing, he risked letting Lyft and Sidecar grow uncontested and undercut him on price. "We were getting so much regulatory heat for the black-car business and it was so clearly legal, that when we saw what I called a regulatory disruption, we didn't think it was going to fly," he told me that year.

The best option seemed to be to watch and wait. Around that time, Zimride co-founder Matt Van Horn, the high-school friend who had accompanied Green to Zimbabwe, ran into Kalanick on a largely empty Muni subway car headed downtown, and he asked him what he thought of Lyft.

"Not legal," Kalanick grumbled, according to Van Horn. "If it's legal, we'll do it too."

That fall, the California Public Utilities Commission seemed to confirm Kalanick's suspicion. It sent cease-and-desist letters to Lyft, Sidecar, and TickenGo, a French company that had just moved to San Francisco and introduced its own ridesharing app for the iPhone.[22] The companies were allowed to operate but ordered into talks with the CPUC, which regulates common carriers like limousines, airport vans, and moving services, as well as the state's public utilities.

To represent its interests in the battle ahead, Lyft hired Susan Kennedy, the assertive and hyperconnected former chief of staff of Governor Arnold Schwarzenegger and, before that, one of the five commissioners of the CPUC. Kennedy knew intimately the byways and internal rivalries of her former agency, located in a building with an elegant circular stone façade at the intersection of McAllister Street and Van Ness Avenue. The entire agency had recently come under intense scrutiny after the 2010 explosion of a gas pipeline in nearby San Bruno that killed eight people, including a CPUC employee and her teenage daughter. She also knew that while the cease-and-desist had emanated from the enforcement division on the second floor, which was led by a brigadier general named Jack Hagan who was known for wearing a firearm in a holster on his ankle, the real policy-making decisions happened on the fifth floor, in the offices of Michael Peevey, her former boss and the CPUC president. The departments on the second and fifth floors, populated by enforcement officers and lawyers, respectively, with widely disparate dispositions and goals, were often fiercely at odds with each other.

Kennedy's easy familiarity with policy makers on the fifth floor likely affected the course of this story.

California was ground zero in the ridesharing movement. It was the first to regulate the ridesharing upstarts, and its deliberations were being closely watched—not only by Travis Kalanick at Uber but by other states who knew the ridesharing phenomenon would be coming to them soon. Nearly the first thing Kennedy did was stride confidently into Peevey's office, where, to her surprise, she found local attorney Jerry Hallisey, who represented Uber. "When are you going to shut these guys down? When?" Hallisey was asking Peevey, who also recalled this conversation.

Kennedy plopped into a chair and listened. After Hallisey left, she started talking into Peevey's ear and didn't stop for weeks. "This is a monumental shift and a brand-new industry," she told him. "This is the cradle and you will either be the guy who was standing in its way and crushed an industry or the guy who facilitated a whole new world."

Their exchange continued over e-mail, which Kennedy later shared with me. She argued that the CPUC needed to start an OIR, a formal rule-making process, to devise guidelines for something that was genuinely new. "The cease and desist approach is the wrong one," she wrote, noting that Lyft and Sidecar did not employ drivers and therefore did not formally fall under his jurisdiction and could fight any order in court. "What problem are you trying to solve with regulation, particularly in a competitive, nascent market? Protect the taxicab industry? Passenger safety? Regulation for regulation's sake? You need to answer this question before you try to shut these services down... Can we talk about this some more before the staff goes off the deep end?"

"You have some good points," Peevey wrote to Kennedy. "Still, I have this nagging fear that the area of ridesharing, aided by technology, will grow and grow and there will be terrible accidents, with the drivers only having minimum insurance coverage."

Peevey was prescient. But Kennedy compared that fear to absurd local efforts to limit the expansion of wireless phone service because people were afraid they might be cut off from emergency 911 services if a cell phone battery died. She noted that Lyft and Sidecar were touting one-million-dollar backup insurance policies, a complement to a driver's personal coverage. And she suggested there was an aura of inevitability around the ridesharing services, linking them to organized carpooling across the Bay Bridge and arguing that "you can't put the genie back in the bottle."

Peevey, in his midseventies, was an economist and lifelong public servant who wore old-fashioned eyeglasses and had a prosthetic nose, a souvenir of his battle with skin cancer. He would leave the CPUC in 2014 under criminal investigation for pressuring PG&E and Southern California Edison, the state's energy utilities, to contribute funds to research groups he supported.[23] (As of late 2016, no charges had been filed.) But Peevey was also defiantly proud of being pro-innovation and as a longtime resident of San Francisco, he had his own personal experiences with the eminent failures of the local taxi industry.

"I used to get into these arguments with the cab guys," Peevey recalled when we spoke at a Starbucks near his Los Angeles home in 2015. " 'You just want to condemn these people but you don't offer anything yourself, you don't want to offer anything new, all you want to do is ring our building with your cabs in protest and honk at us.' "

If Peevey ever considered shutting down Lyft and Sidecar, Kennedy quickly turned him around. That fall, he instructed Marzia Zafar, his director of policy, to let the ridesharing companies operate but to figure out a way to protect the safety of riders.

Zafar, who ran the subsequent rule-making process and would write the eventual regulations, cut an unusual figure for a regulator. She was an Afghani immigrant with a Mohawk who had moved to the United States as a child and had once driven a cab herself for her uncle's taxi company in San Bernardino County. Zafar and her colleagues invited representatives from all the various interested parties in for conversations, and there she got an education in the profound differences in these emerging markets.

The taxi and limo companies all arrived separately, stating their gripes against Lyft and Sidecar but also Uber and, comically, one another, based on decades of simmering hatreds. Travis Kalanick also visited the CPUC conference room on the fifth floor that fall, with Hallisey, his attorney, and general counsel Salle Yoo. He made a lasting impression. "It was the strangest thing, I still remember," Zafar says of the meeting. "He basically turned his chair the other way, toward the wall. His back was facing us, very deliberately." Zafar recalls Kalanick's first words as being "Why don't you take Lyft out of the market? They are not complying with your regulations!"

Zafar's colleague Paul Clanon, the CPUC's executive director, later reflected, "The guy's a jerk, but I have to say I kind of have a soft spot for him. Maybe the way you build organizations as successful as Uber is by not giving a fuck what regulators think of you."

Logan Green and John Zimmer, escorted by Susan Kennedy,

also made their way up to the fifth floor. They were earnest and personable, explaining with their usual missionary zeal their goal of filling empty seats in cars.

"They always came across as choir boys," says Kennedy. The CPUC "had to trust somebody when writing new laws. You can't write in a vacuum, you have to listen to industry." She speculates that if Uber had been the original advocate for ridesharing, there might have been provisions like drug testing of new drivers, which would have slowed down sign-ups and impeded the growth of the industry. "I wonder if Uber realized what they did for them," Kennedy says.

The CPUC reached a consent decree with Zimride and Sidecar in January 2013.[24] The companies agreed to comply with basic safety requirements, like requiring drivers to show proof of insurance and checking their backgrounds for criminal history, which they were already doing. The decree also required them to check drivers' DMV records for driving infractions, which they were not. Then they were allowed to operate pending the creation of a new set of rules, which were slated to be formulated after a comment period and a public hearing that spring.

A few weeks later, Lyft expanded into Los Angeles and Sidecar moved more aggressively into L.A., Philadelphia, Boston, Chicago, Austin, Brooklyn, and DC.

The ridesharing wars had begun.

Travis Kalanick had watched, waited, and even quietly agitated for Lyft and Sidecar to be shut down. Instead, they spread, undercutting Uber's prices. Now that their approach had been sanctioned, Kalanick had no choice but to drop his opposition and join them. In January 2013, Uber signed the same consent decree with the CPUC and turned UberX into a ridesharing service in California, inviting nearly anyone with a driver's license and proof of insurance, not just professional drivers, to open his or her car to paying riders.[25]

Kalanick then announced his broader intentions to compete nationally with the ridesharing companies with a seminal white paper, posted to the Uber website and grandiosely titled "Principled Innovation: Addressing the Regulatory Ambiguity Around Ridesharing Apps."

"Over the last year we've stayed out of the ridesharing fray due to perceived regulatory risk and watched two competitors roll out in a few cities in which we already operate, without nearly the same level of constraints or costs, offering a far cheaper product," he wrote. "In the face of this challenge, Uber could have chosen to do nothing. We could have chosen to use regulation to thwart our competitors. Instead, we chose the path that reflects our company's core: we chose to compete." Uber, he wrote, would add ridesharing to UberX nationwide and roll it out in cities where there was tacit approval, regulatory ambiguity, or an absence of enforcement. Drivers would have to undergo online background checks and would be covered by a million-dollar liability policy to be held by a wholly owned Uber subsidiary, which was named Rasier—German for "shave."

Uber, in other words, was coming after the mustache.

There was no shortage of animosity between the companies and their willful execs. Around that time, Kalanick and John Zimmer got into a heated and puerile battle on Twitter, accusing each other of having inadequate insurance and ineffective background checks; "@Johnzimmer, you've got a lot of catching up to do...#clone," wrote Kalanick, angling for the last word.[26]

But by April they were effectively on the same side at the CPUC's public workshop, a meeting intended to gather input for the forthcoming set of new laws to regulate ridesharing. The hearings, held on April 10 and April 11, 2013, in the auditorium in the CPUC building and open to the public, were just the kind of circus that would be replicated, with slight variations, in countless cities, states, and countries all over the world over the next few years. Angry taxi drivers, their unions, execs from Uber and the other ridesharing companies, and interest groups representing the disabled and the blind all crammed into the auditorium at 505 Van Ness to loudly articulate their concerns.

"People don't like to talk about the fact that this competition will kill our taxi industry," railed one of the first speakers, Christiane Hayashi, head of the San Francisco MTA and Uber's first regulatory foe, to applause from the gathered cabbies. "But when this unregulated and illegal competition has devastated the landscape, no one will be left to provide universally accessible door to door transportation services to our residents. Should they be regulated as taxicabs? Yes!"

Hayashi had previously taken two Lyft rides, paid nothing of the suggested donation, and then marveled that Lyft drivers would no longer pick her up. When she asked John Zimmer why over breakfast one morning, he looked up her ride history on his phone and noted that she hadn't paid anything. Hayashi was outraged by what she viewed as a privacy violation.

With limited success, Marzia Zafar tried to keep the proceedings moving and civilized. The ridesharing companies testified, one after the other, often over the jeers and taunts of the assembled taxi drivers. After the debacle in DC, Uber's lawyers kept Kalanick far from the proceedings; instead, Ilya Abyzov, general manager of the business in San Francisco, took the podium and insisted that Uber was only a software company. "Our offices have programmers, not drivers," he said. "Uber is agnostic about ridesharing. Whatever the decision is, we will follow it." When Zafar opened the session to questions, an immigrant driver stood up and cut through Uber's carefully crafted distinctions. "Sooner or later, you will have to face the issue that you are a car service," he yelled.

The testimony of Lyft's attorney Kristin Sverchek grew even more heated. When the discussion turned to insurance, one driver, a local medallion holder, started showering her with profanity. "Hold on, hold on! He just referred to me as a dumb bitch," Sverchek protested at the podium. "I think that's completely inappropriate." Zafar agreed and threw the driver out of the auditorium.

The decision by the five PUC commissioners on the ridesharing companies was ultimately unanimous. Under Michael Peevey's influential direction, and with letters of support from Mayor Ed Lee in San Francisco and Mayor Eric Garcetti in Los Angeles, Peevey and the four other commissioners voted to formally legalize ridesharing, classified the firms as "transportation network companies," and said they would revisit the ruling in a year. The new rules required the companies to, among other things, report the average number of hours and miles each driver spent on the road every year—a requirement Uber would subsequently ignore, racking up millions in fines.[27] It also reiterated that the companies were required to hold a million dollars in supplemental insurance to cover drivers, but only while passengers were in their car—a provision that was soon shown to be tragically inadequate.[28]

Nevertheless, the ruling legitimized the TNCs and gave them ammunition for coming legal fights in other states and countries. And it tilted the playing field back in the favor of Uber, which had more resources in more cities. Now Ryan Graves, Austin Geidt, and their launch teams could analyze the market in each new city and decide whether and when to launch Uber Black, UberX, or Uber Taxi.

The ruling had some unintended consequences. After Marzia Zafar confessed to drafting the new regulations, her uncle, the taxi-company owner in San Bernardino, didn't speak to her for a year.

It also put pressure on some of Uber's first and biggest fans—its black-car drivers. Sofiane Ouali, driver of the white 2003 Lincoln known around San Francisco as the unicorn, had spent his savings to lease half a dozen cars for a black-car business he was running on Uber's platform. His company, Global Way Limousine, flourished for a year, at one point with sixteen drivers taking shifts. But when ridesharing took off, Ouali knew trouble was ahead. Fares were coming down and drivers no longer had any reason to split their commissions with a fleet owner—they could just drive their own cars and work directly for Uber. "I never got angry about it," says Ouali, who returned his extra cars but kept driving. "I understood that Uber couldn't risk its business."

In a kind of cosmic irony, the unicorn itself was totaled in an accident on St. Patrick's Day when a drunk driver ran a red light (no one was seriously injured). Ouali decided not to have it repaired. "I ended up thinking maybe it was the right decision," he says. "Unicorns work like this. They disappear and maybe magically reappear someday."

Uber had survived its most serious challenge yet, pivoted (albeit reluctantly at first) into what would prove to be a much larger business, and shown itself to be a flexible player unwilling to surrender leadership in the field of transportation apps for smartphones. It had Sidecar and Lyft to thank, which Kalanick was inclined to admit when he was in a charitable mood. "The one area where they brought some thunder was on that regulatory piece," he told me in 2014. "I look at entrepreneurism as risk arbitrage. You are basically looking at risk and saying, 'I think people are misunderstanding it and I'm going to go after it.'"

Kalanick had been among those who had overestimated the risk, misplayed it, and been bruised by it. Uber had waited on the sidelines of ridesharing for seven months, and in that time new rivals had gained critical momentum. What Lyft and Sidecar did was ambitious, he conceded, vowing: "We are not going to let this happen ever again."

Sidecar expanded too aggressively, and its drivers' cars were impounded in New York, Austin, and Philadelphia.[29] Lyft, more careful, was building a distinctive brand. It would become Uber's most tenacious competitor in the United States.

The lessons of the past year now seemed obvious. Moving cautiously and playing by the rules had proven to be a costly mistake. People around the world wanted these new transportation options, and according to Travis's Law, their fervor could provide the political cover to fuel a rapid expansion. If the taxi lobby and their political surrogates didn't want to let the future unfold, well, he had seen that movie before, with the music industry during his file-sharing days. There was no point trying to negotiate with them. To maintain Uber's position in the vanguard of upstarts changing transportation, Kalanick, already aggressive and determined, was going to have to be even more aggressive and determined—and even, perhaps, a little bit ruthless.

It was an attitude that would change the world's perception of Uber. And despite earnest protests to the contrary, it would rub off on one of Uber's fellow upstarts—Airbnb.

CHAPTER 9

TOO BIG TO REGULATE

Airbnb's Fight in New York City

> Who are you renting my apartment to? What the hell is going on there?
>
> —Landlord Abe Carrey to tenant Nigel Warren, September 2012

Belinda Johnson, Airbnb's first in-house attorney, started visiting lawmakers in the spring of 2012. The rapidly growing startup was pitching its service as a financial opportunity for hosts and an economic boon to communities trying to draw in more tourist traffic. But neighborhood groups and some regulators weren't so sure, viewing it more as a way for disreputable landlords to evict tenants and convert their buildings into condos or illegal hotels. Johnson's job was to change their minds.

A polished executive in her forties, Johnson had worked closely with regulators and law enforcement officers for years in her previous job at Yahoo, dealing with issues like privacy and online child safety. Transparency, collaboration, and compromise—these, she vowed, would also be the operating principles for the legal and public-policy teams at Airbnb.

But her first round of meetings on behalf of Airbnb did not go well. Most legislators either hadn't heard of the home-sharing site or just didn't understand it. Do hosts leave their houses? Do they actually sleep under the same roof with total strangers? On a trip to New York City, one official put his thumb and forefinger up to his mouth and theatrically inhaled—suggesting Johnson and her colleagues were smoking pot if they thought such a thing would ever catch on. "We were still so small that we were under everybody's radar," she told me. "It seemed a little hippie-dippy. So yeah, we had to work on our story."

The gesture would turn out to be one of the more tolerant responses from New York State regulators regarding the emerging home-sharing site.

Johnson started her career in the 1990s as a junior attorney, spending six years at a series of stultifying Dallas law firms. One day she met a high-profile local entrepreneur named Mark Cuban at her gym and asked how she could help with his thirty-person internet radio startup, called AudioNet. It was 1996, and AudioNet was operating from a three-thousand-square-foot warehouse in downtown Dallas that had mice in the bathroom and not enough chairs for employees. Johnson joined as its first lawyer, helping the renamed broadcast.com to convince Texas colleges to put their sports broadcasts online and to navigate what were then completely unexplored aspects of copyright law.

Cuban, the future owner of the Dallas Mavericks and mainstay of the TV show *Shark Tank,* was a visionary; he predicted that one day sports and other programming would be streamed online, breaking conventional television's stranglehold on the media business. Cuban and his co-founder, Todd Wagner, were ahead of their time, and there seemed little chance back then that the company would ever get to the promised land of profitability. Nevertheless, in 1999, Yahoo, high on the financial narcotic known at the time as internet stock, bought broadcast.com with its own overvalued shares in a deal worth $5.7 billion.

Johnson moved to San Francisco and spent the next decade as Yahoo's deputy general counsel. She worked for four different CEOs at the increasingly beleaguered web portal, and by 2011, wanted to believe in another corporate cause. That's when she started reading in the tech press about Airbnb.

Impressed by the startup's gathering momentum, Johnson stealthily orchestrated her own hiring. Instead of sending an unsolicited e-mail to Brian Chesky, she asked well-known Silicon Valley investor Ron Conway to broker a connection. Instead of campaigning for a full-time job, she first offered her services as a consultant. She then won Chesky's trust, in part by enthusiastically embracing the company's sense of its own virtue, its near religious certainty of its position in the vanguard of a historic new sharing economy that could change the world.

The Airbnb that Johnson joined full-time as general counsel in December 2011 was almost irrationally consumed with its own identity. Employees avidly read and discussed the newly published book *What's Mine Is Yours: The Rise of Collaborative Consumption,* by Rachel Botsman and Roo Rogers, which theorized that the twenty-first century was not about individual purchasing habits or the conventional idea of owning things but about internet communities, online reputations, and the efficient sharing of underutilized resources.

Execs spent months hashing out the company's six core values—"Be a host," "Every frame matters," "Simplify," "Embrace the adventure," "Be a 'cereal' entrepreneur," and "Champion the mission." The last one proclaimed awkwardly: "The mission is to live in this world where one day you can feel like you're home anywhere, and not in a home, but truly home, where you belong."

Chesky introduced these values to employees at a company offsite held at the Sonoma estate of sculptors Lucia Eames and Llisa Demetrios, daughter and granddaughter of the famed furniture designer Charles Eames, whom Chesky had idolized in design school. The six values would be used to guide hiring decisions and employee performance reviews and to illustrate Airbnb's ideas about itself to the world.

As part of this protracted process of self-reflection, the company executives also debated whether to move into other sharing-economy businesses, such as the rentals of cars and office space between customers. Ultimately, Chesky decided to hold off on that type of expansion and double down on home-sharing by studying and fine-tuning the process of renting and hosting on the site. Obsessed with all things related to Disney, he dubbed this internal review Snow White, after the iconic movie, and hired a computer animator from local movie studio Pixar to storyboard the "emotional moments" of Airbnb customers.[1] The panels—which illustrated the Airbnb experience from the perspective of hosts thinking about what they could do with extra income and guests excitedly spreading the word about the service—were mounted on the walls of the main conference room, dubbed Air Crew, in the Rhode Island Street office.

Fresh from scattered, chaotic Yahoo, Belinda Johnson was impressed. "I loved the creativity," she says. "When you have so much opportunity, being able to say no is going to distinguish a company that is going to do well from one that might lose its way."

Johnson's formal title at Airbnb was general counsel, but as Chesky's first significant senior hire, she became more of a consigliere. She helped Chesky recruit his original chief financial officer, Andrew Swain, who came from the accounting software maker Intuit, as well as Mike Curtis, vice president of engineering from Facebook, who could help Nate Blecharczyk manage a large engineering team and a global, rapidly scaling website. Chesky trusted Johnson and befriended her; they twice attended the annual Burning Man festival in the Nevada desert together with groups of friends and colleagues and he said they spoke "every day, multiple times a day."[2]

Chesky had good reason to make a veteran attorney his first major outside hire. The company had mounting regulatory challenges around the world, where its fierce sense of its own righteous mission was clashing with an increasingly hostile reception in cities like San Francisco, Barcelona, Amsterdam, and particularly New York City, its largest market at the time.

Throughout 2012, Johnson watched Travis Kalanick's fiery battles in DC, San Francisco, and elsewhere, and believed Airbnb had to do things differently than Uber.

She talked about ethereal concepts like Airbnb's "regulatory brand" and stated things like "It has to be authentic to who the company is," reflecting the fact that Airbnb "operates in a principled way." Making the rounds of influential lawmakers and talking to them face to face was the first step. "We wanted to build a positive kind of credibility with cities," she insists. "That is just better in the long run but most importantly it was authentic to who our founders are."

But just a year after she was accused of being high for thinking that such a business scheme could work, another reaction by a New York politician showed that the emerging dynamic was considerably more combative. By then she had hired David Hantman, another Yahoo refugee, to lead Airbnb's public-policy team. Hantman and his colleagues were canvassing New York City, trying to spread the gospel of Airbnb's positive impact on the community, when they encountered Liz Krueger, a fiery state senator from Manhattan who for years had battled illegal hotels in the city. Krueger's office, it turned out, was besieged by complaints about Airbnb from angry neighbors and from hosts who had listed their apartments as short-term rentals on the internet and who, to their surprise, then faced eviction notices from landlords, since many New York City leases expressly forbade subletting of any kind.

Krueger didn't seem to believe in the mission of Airbnb, its "regulatory brand," its corporate values, Snow White, or the immaculate hearts of the three founders. She had a withering assessment for Hantman and his crew: "I have never dealt with a company as disingenuous as Airbnb has been over and over and over again," she said.

* * *

To understand why Airbnb's public reception among some lawmakers contrasted so dramatically with its virtuous view of itself, we must rewind this narrative, go from the stylish offices on Rhode Island Street, back through the garage on Tenth, all the way to the original apartment on 19 Rausch. It is once more early 2009, two years before Belinda Johnson joins the company, and Airbnb is again Airbedandbreakfast.com.

Brian Chesky and Joe Gebbia, ambitious and battle-scarred from a year in the startup trenches, were at the Y Combinator startup school when they received an e-mail from a part-time actor, famous New York party planner, and residential landlord whose activities would set the unfortunate tone for Airbnb's future path in New York City. His name was Robert "Toshi" Chan.

A native of San Francisco whose parents emigrated from China, Chan attended Columbia University as a math major and then made millions on Wall Street trading government securities at Citibank. But after seven years, he found that lucrative life too constraining and anonymous. In an act of reinvention possible only in New York, Chan dropped his given name, Robert, in favor of Toshi (the name of the most popular boy in his high-school class) and launched a career as an actor. "With my 'Masters of the Universe' ego at twenty-five years old, I thought, *If I can trade billions, I don't see how hard it can be for me to win an Academy Award,*" he told me.

The charismatic, relentlessly self-promoting Chan won bit parts in *Law and Order, Late Night with Conan O'Brien,* and the Martin Scorsese film *The Departed,* where he played a jittery mafioso. But mostly he made his name by holding famously over-the-top parties a few times a year, with $1,500 ticket prices and topless Toshettes covered in body paint.[3] (*AM New York:* "He's King of the City That Never Sleeps.")[4] He used his Wall Street winnings to buy a four-story former yeshiva on a quiet street in south Williamsburg and renovated the entire structure, adding a lavishly appointed two-floor penthouse with eighteen-foot-high ceilings.

The sequence of events that would affect New York State housing law and the future of Airbnb started to unfold in 2007. With acting jobs sporadic and finding it increasingly difficult to obtain liquor licenses and venues for his parties, Chan was basically unemployed. His then fiancée, Cha Chang, recalls that around this time Chan rented out one of the guest rooms in his penthouse suite for a few weeks to a friend from Sweden. When the friend left, Chan posted the room on Craigslist for one hundred and fifty dollars a night.

To someone as clever and opportunistic as Toshi Chan, the favorable economics of short-term rentals must have been quickly evident. He could rent apartments in his building for fifteen hundred a month; on the internet, he could charge tourists one hundred fifty a night, and if he rented a room out twenty days a month, he'd earn three thousand from one room alone. Soon after, Chan began listing apartments in the six-floor building next door, which he had leased on favorable terms from the landlord. Tourists, fresh from the airport, started pouring into their home to pick up their keys. Cha Chang devised a breakfast menu, charging guests five dollars for eggs or pointing them to nearby diners.

The woes of the New York City real estate market started to deepen in 2008. Landlords had open apartments and plenty of tenants who couldn't pay their bills. This was Chan's big break. He signed annual leases on a dozen cheap two-bedroom apartments around the corner and posted those to Craigslist too. When the online bulletin board proved too cumbersome to use for multiple listings, Chan expanded, creating his own website, HotelToshi.com, and turned to tourism services like FeelNYC.com, popular in Europe, and Roomorama, a New York–focused apartment-rental site that had opened that year. When Cha Chang read an article about AirBed & Breakfast, she added that to their staple of listing sites too.

In early 2009, while Airbnb was still in Y Combinator, Chan and his assistants started corresponding with Chesky. The young CEO advised Chan that he could pay twenty-nine dollars a year and upgrade to a premium membership, a short-lived offering that allowed hosts to list properties that were priced at over three hundred dollars a night. "Many of our premium listers are our best hosts," Chesky wrote in an e-mail that February that Chan later shared with me. "I would be happy to talk to you about this, and arrange something that works for you. How many listings are you looking to post?" Despite Airbnb's later attempts to distance itself from hosts who posted multiple listings, back then Chesky and his co-founders welcomed them.

Chan remembers Chesky and Joe Gebbia staying overnight at one of his apartments in Brooklyn and on another occasion having dinner at a sushi restaurant in Tribeca with Chesky, Gebbia, and investor Greg McAdoo, where they discussed topics like how to streamline the check-in process. In June 2009, young Airbnb had only eight hundred listings in New York City, and Toshi Chan was providing at least fifty of them.[5]

As the financial crisis worsened, Chan accelerated his plans. He found a co-investor and signed leases for some two hundred other apartments across Brooklyn and on Manhattan's Upper West Side. He even set up a tent and a queen bed on the roof of his penthouse and rented it out on Airbnb for a hundred dollars a night. "He lets guests use the bathroom in his apartment," wrote a *Daily News* reporter who stayed there and wrote about it that month.[6]

As the business grew, Chan moved his office out of his home and into the basement of a nearby building in Williamsburg, where he rented about half of the thirty-five units. Instead of having tourists coming there to pick up their keys, the Hotel Toshi sent bike messengers and eventually a van adorned with the cartoon logo of him, his head resting on a pillow, to meet guests at the apartments. Coordinating it all was utter chaos, recalls Cha Chang, who had become a Hotel Toshi employee. Orchestrating check-ins and obtaining enough clean sheets every day was a particularly nightmarish task. Perhaps the worst part was the incessant, shrieking phone calls from the permanent residents of the buildings, who were understandably livid about the nonstop tourist traffic and late-night partying by guests.

At its peak, the Hotel Toshi had more than a hundred employees. They all suspected the operation was of dubious legality and lived in a constant state of panic that the company would get shut down and they would be arrested. According to Cha Chang, Toshi paid off two separate individuals who blackmailed him, threatening to report the Hotel Toshi to city authorities. "It was a business cost," says Chan when asked about the blackmail attempts. "The consequences of not paying would have been much worse."

But it turns out the city was already watching the Hotel Toshi and similar operators. For the previous five years, well before the founding of Airbnb, the administration of Mayor Michael Bloomberg had been looking for a way to address the scourge of avaricious landlords harassing and evicting low-income residents in order to convert apartments into illegal hotels or condos and meet the demand for cheap rooms with kitchens close to major tourist areas.

Around 2006, housing advocates and elected officials across a range of city and state agencies had begun meeting to discuss the issue. The task force eventually proposed rules, amendments to a 1968 ordinance called the Multiple Dwelling Law, that stipulated that permanent residents of an apartment building could not rent their homes to guests for any amount of time less than thirty days. The law, which would effectively make very short-term home-sharing and subletting illegal in New York City, came up for a vote in the state legislature in the summer of 2010—just as complaints against the Hotel Toshi and Airbnb were reaching a fever pitch.

Chesky says he heard about the law only days before the vote. In response, he hired Airbnb's first lobbyist, a well-known Albany lawyer named Emily Giske, who started visiting state lawmakers. Meanwhile, Toshi Chan met some of New York's biggest landlords in the downtown offices of the law firm Fried Frank to discuss the options. They would eventually help create an advocacy group called Save Sublets to try to marshal opposition to the city's plan.

On July 21, the landlords and a group of apartment-listing websites organized a protest at city hall. Airbnb co-founder Joe Gebbia flew to New York to attend the event and Tweeted to his followers to join him and "save sublets."[7] Chan remembers going to the protest with Gebbia, waving signs and asking people to put their names on petitions. "Toshi, maybe it's better if you're not out in front," Gebbia told him, according to Chan. "They kind of hate you."

Chan recalls all of this from the penthouse of the Flatiron Hotel on the corner of Twenty-Sixth and Broadway in Manhattan. He invested in the boutique hotel in 2011 and became principal owner in 2014. Off the lobby, there's a nightclub called Toshi's Living Room, where guests can catch a jazz quartet on most nights. In the penthouse, there's a party venue and an outside deck adorned with Toshi's cartoon visage. As he reminisces, Chan reclines on a couch and strokes Ponzu, his white Maltese Yorkie. "I went from making five thousand a month to twelve million a month in just a couple of years," he says. "It was crazy."

But it wasn't sustainable. "Hotel Toshi was so toxic. Every neighbor hated me. I was like the anti-Christ, I was worse than Hitler in their eyes," he says, a tad mischievously. "Landlords were kicking people out to rent out their apartments to me. That was not good. In retrospect, there's a certain social responsibility you learn as you get older."

Chan didn't get enlightenment all at once. The Multiple Dwelling Law passed and was signed by Governor David A. Paterson but it wasn't officially implemented until May 2011.[8] Instead of shutting down, Chan paid his fines and changed his company's name to Smart Apartments, in part because the Hotel Toshi brand had attracted too many scathing reviews. He tried to keep posting on Airbnb but says that after the law passed, the company canceled his listings. "They dropped me like a hot potato," Chan says. "I understood. It was smart."

But for Chesky and Gebbia, the realization that they had associated with an unsavory character may have come too late. In the eyes of law enforcement and the media, the brands of Toshi and Airbnb were closely intertwined. In October 2011, Chan was sued by the Office of Special Enforcement, a division of the mayor's office tasked with solving quality-of-life problems like illegal hotels.[9] The city charged him for a litany of fire-code violations as well as for operating unsafe and illegal accommodations. The lawsuit, Chan says, "had the might of Thor's hammer." With the hammer coming down on him, he eventually settled the case for a million dollars and shuttered Smart Apartments. "Infamous Airbnb Hotelier Toshi to Pay $1 Million to NYC," wrote the *New York Observer*.[10]

Brian Chesky and Joe Gebbia had been friendly with Toshi Chan, had stayed in one of his apartments, dined with him, and facilitated his business. So they couldn't move on from the association quite so easily. Airbnb would now have to live with the consequences of a stifling law in New York City that Toshi Chan and others like him had helped to produce.

By 2012, Belinda Johnson and her colleagues knew they faced a daunting obstacle—many of Airbnb's best customers in New York City were essentially behaving illegally. Even worse, Airbnb could do little to change what it felt was an exceedingly restrictive law. When the ordinance passed, Gebbia had tried to summon supporters to join him at the city hall protest, but Airbnb was so small it didn't have a real community to advocate on its behalf. Travis's Law didn't yet apply.

There was an obvious solution, even though it meant putting its own hosts in legal jeopardy; Airbnb didn't tell hosts they were breaking the law. Then the company tried to grow in New York and ingratiate itself to its users, with an aim of getting them to influence city officials on its behalf. "We needed to get big enough to win," says a lawyer who worked with the company at the time.

Airbnb believed it had assurances from city and state officials that, despite its broad wording, the Multiple Dwelling Law applied only to residents who left the premises and rented their homes to tourists for under thirty days. This type of rental, an unhosted stay, amounted to an illegal short-term sublet. According to New York State officials, these hosts were acting like they were hotel owners, not the hosts of a bed-and-breakfast. People who stayed on the premises and rented out a spare room or couch—like the much-touted Airbnb user who was using the site to meet people and make ends meet—were behaving perfectly within the bounds of the law, the company believed. Still, there were ominous signs that the city wanted to curtail the activity altogether. In 2012, with a steady stream of angry complaints from neighbors about Airbnb and other short-term rental sites pouring into government offices, the city council increased fines to twenty-five thousand dollars for repeat violations of the Multiple Dwelling Law, up from the original penalties of less than three thousand dollars.[11]

Airbnb was facing a wave of hostility from lawmakers and couldn't seem to find a way to stand up for itself or its host community. And then Nigel Warren, a thirty-year-old resident of the East Village, came home one afternoon in September 2012 and received an irate phone call from his normally placid landlord, Abe Carrey, an elderly man from Queens. Abe yelled, "Who are you renting my apartment to? What the hell is going on there?" Warren's stomach dropped.

Warren is a hip, soft-spoken web designer—your typical East Village denizen, in other words. He had used Airbnb just three times to rent out his room while traveling over the previous year.

His roommate, Julia, had rented out her room once. Their experiences were positive and profitable, earning them a little over a hundred dollars a night, a modest contribution to their three-thousand-dollar-a-month rent for a two-bedroom, one-bath sixth-floor walk-up. A week before the call from Abe, Warren had gone away to Colorado with friends for five days and had used Airbnb to rent out his room to a tourist from Russia who spoke little English. Julia was there and things had gone smoothly, apart from the guest's vague account of running into some police officers in the hallway. "There were no horror stories," Warren says. "Everything was perfectly fine."

Except it wasn't. The Office of Special Enforcement had been tipped off, perhaps by annoyed neighbors, that Warren and his roommate were subletting space in their apartment. (A former member of the department later told me they had reason to believe—inaccurately, it turned out—that Nigel Warren was essentially another Toshi Chan.) Building inspectors showed up while Julia was gone, questioned the guest from Russia in the hallway, noted a few safety violations in the building, and left. Then they mailed a notice to Abe's home in Queens alleging that the tenants in apartment 5G were running an illegal transient hotel and violating safety codes. The potential fines exceeded forty thousand dollars. Warren assured the irate Abe that he would accept responsibility and deal with it, but at the time he was freelancing and without a steady income. "That began many months of stress," he says.

Much of his fury was directed at Airbnb. When Nigel Warren sat down to research the rules he had allegedly broken, he found articles on the 2010 law, which the website hadn't warned its customers about. There was some fine print in its twelve-thousand-plus-word terms and conditions that advised hosts they were responsible for understanding local regulations, but Warren of course hadn't read that lengthy document.

Warren's sister recommended he get a lawyer. The attorney

concluded that Warren was probably safe because Julia had been present during the guest's stay. Nevertheless, at $415 an hour, the lawyer's fees only added to his financial misery. To save money, Warren decided to argue his own case. The first hearing was canceled due to Hurricane Sandy, which resulted in a long delay that led him to mistakenly believe he was in the clear. Then he was called back to appear in housing court. The city was throwing the book at Nigel Warren, hoping to set an example to curb usage of Airbnb.

At the time, Airbnb was enjoying a wave of positive press for new features and another $200 million in funding led by venture capitalist Peter Thiel, a PayPal co-founder and early Facebook investor. Warren seethed that the company was enjoying all this adulation while embroiling hosts like himself in legal trouble. Finally, he decided to take action. He did two things. First, he e-mailed Airbnb, complaining in part, "This entire situation came as a complete surprise. I had no idea that being an Airbnb host is illegal in most locations in New York City."

The company responded five days later. "I am sorry to hear that you are gong [*sic*] through a stressful situation," e-mailed a customer-service agent, Maria C. "We encourage you to familiarize yourself and comply with any and all lease, rental, or co-op agreements, as well as applicable local, state, national and international regulations. When renting, special local and state taxes may also be required. It is the responsibility of hosts to comply with all such regulations and taxes." Maria C. unhelpfully concluded, "We're happy to see you are staying informed and aware!"

The second thing Warren did produced better results. A friend introduced him to Ron Lieber, the "Your Money" columnist for the *New York Times*. Warren told Lieber his story, and "A Warning for Hosts of Airbnb Travelers" was published on November 30, 2012. "Many people believe that living on the web grants them membership in an exalted class to which old laws cannot possibly apply,"

Lieber wrote. "This sort of arrogance takes your breath away, until you realize just how brilliant a corporate strategy it is. If you stopped to reckon with every 80-year-old zoning law or tried to change the ones that you knew your customers would violate, you'd never even open for business."[12]

A few hours after Lieber's story was published, Warren received a more contrite call from an Airbnb customer-service representative. And when he showed up for his next hearing, once again without his pricey lawyer but accompanied by a National Public Radio reporter who was chronicling his saga, Warren was surprised to run into David Hantman and two lawyers. They told Warren that, like any internet company, Airbnb couldn't provide legal advice or financial support, because that might set a precedent that would make the company responsible to assist everyone who used the service. But they had filed an independent brief on Warren's behalf and wanted to observe the proceedings.

The case was a critical one for Airbnb, which believed its users in New York City were at least authorized to share their homes if they were present. A bad decision "could have set a terrible precedent. It was a big problem if the law changed because of some ruling," a lawyer involved in the proceedings told me.

After more preliminary hearings and delays, Warren's case was scheduled for May 2013. Still trying to save money, he represented himself in the hearing room of the Environmental Control Board, a drab DMV-like office on the tenth floor of a city building in lower Manhattan. With Airbnb's fate in New York City hanging in the balance, Warren botched legal procedures, mishandled cross-examinations of witnesses (including the building inspector who had confronted the Russian guest), and was constantly asked to rephrase questions—all while Airbnb's high-priced attorneys sat quietly in the audience. "I was in so far over my head, it was just ridiculous," Warren says.

Five days after the hearing, Warren got a phone call from the court: the judge had dismissed the safety-code violations and struck down the charge that Warren was running a transient hotel. But in a curious bit of legal reasoning, he ruled that Warren had indeed broken the law because the Russian tourist and Julia were "strangers," and the tourist did not have access to every single space in the apartment—in this case, Julia's bedroom. Thus the tourist was not technically "living within the household of the permanent occupant."[13] The resulting fine on the infraction was $2,400. Still, Warren considered the ordeal over and happily paid the penalty. "That was a victory in my book," he says.

But it wasn't a victory for Airbnb. If Nigel Warren was breaking the law, then so was every Airbnb host in New York City. If that was the case, the company had a big problem; it had no legal business there and it would never get big enough to change the rules.

Inside the company, an intense debate erupted. Belinda Johnson and David Hantman believed that Airbnb couldn't let the precedent stand and that it would be a terrible signal to other cities that were considering slapping limits on the fast-growing sharing economy. Other attorneys advising the company worried that if Airbnb appealed, it might be compelled to get involved in other cases involving its hosts.

Chesky made the final call: of course Airbnb should stand up for its hosts. "We needed to advocate for our community," says Belinda Johnson. "It was clear to us that this was the wrong interpretation of the law." Airbnb then hired New York City law firm Gibson Dunn to appeal Nigel Warren's case. As it had with EJ, Airbnb was taking another critical step away from being an aloof, neutral platform and toward a service that advocated for its hosts and was willing to walk in their shoes.

The appeals process took another three months. With serious legal muscle now on Airbnb's side, the Environmental Control Board found that the Multiple Dwelling Law did not, in fact, require a personal relationship between a short-term guest and a permanent resident. The case was dropped and Warren was finally exonerated. "This decision was a victory for the sharing economy and the countless New Yorkers who make the Airbnb community vibrant and strong," wrote Hantman afterward on Airbnb's public-policy blog.[14] Tech-news sites like *TechCrunch* and the *Verge* celebrated the victory.

Perhaps the only person who wasn't celebrating was Nigel Warren himself. "I was happy but not grateful," he says, recalling the whole strange saga in a quiet conference room in the Brooklyn offices of Kickstarter, the crowd-funding website where he has worked as a product manager since 2014. I asked him if he thought Airbnb behaved honorably in his case. "I don't think honor really came into it," he says. "There are certain companies that at certain times act with honor outside the bounds of what the marketplace demands. In this case, I think their actions were purely pragmatic."

Airbnb "went up to the line of what they needed to do. I don't resent them for it. It was clear what the equation was. They had to protect their business in New York."

A few days after the verdict, in a triumphant essay on the Airbnb website entitled "Who We Are, What We Stand For," Chesky used the Warren victory to issue a battle cry, presenting the company's righteous view of itself. "We all agree that illegal hotels are bad for New York, but that is not our community," he wrote underneath a photograph of a group of young people staring over the East River and into a setting sun.[15] "Our community is made up of thousands of amazing people with kind hearts.

"We imagine a more accessible New York that even more people can afford to visit, where extra space in people's homes will not go to waste, and where millions of visitors patronize neighborhood small businesses across all five boroughs," Chesky wrote. "This will be a city where tens of thousands of jobs for people like photographers, tour guides, and chefs will be created to support this thriving new ecosystem."

He added that he wanted to help New York City collect and pay hotel taxes on Airbnb rentals and that he was eager to help the city root out the bad actors causing disturbances in residential neighborhoods, which he proposed to do primarily by setting up a 24/7 hotline to field complaints.

Among city and state officials, the screed went over poorly. Liz Krueger, the New York senator who slammed Airbnb as disingenuous, says her office at the time was deluged with complaints from constituents. With New York real estate starting to recover from the recession, landlords were leaping at excuses to free up rent-controlled apartments and lease them again at the higher market rates.

Krueger met with Airbnb representatives and urged them to warn hosts on the site, with clearly visible language, that they might be violating both state law and their leases. Airbnb, she says, responded with a rotating series of explanations of why that was too complex or how it exposed the company to legal liability. (The site was still not adequately warning New York customers a year later, according to a review by *Gawker*.)[16] Krueger, a lifelong New York Democrat with a dry wit and a dim view of Silicon Valley startups seeking to play by their own rules, figured there was a simpler explanation: Airbnb didn't want to curtail its fast-growing business in the city. And she laughed at the ridiculousness of the neighborhood hotline and the idea that the California company might be able to respond meaningfully when complaints came in during the middle of the night or weekends.

Meanwhile, lawyers for the state attorney general, Eric Schneiderman, the top law enforcement officer in the state, were inclined to agree. They felt that despite its proclaimed intentions to aid the city, Airbnb was actually resisting requests to combat illegal hoteliers and hadn't earnestly pursued efforts to collect the required 14.75 percent hospitality taxes.[17]

Though it hadn't been made public yet, the attorney general was calling Chesky's hand. In August 2013, Schneiderman subpoenaed

Airbnb to provide the names, addresses, and contact information for all Airbnb hosts in New York State, as well as the dates and duration of guest stays and the fees they earned since early 2010. In private meetings, Airbnb refused, and then, after the Nigel Warren verdict, Schneiderman reissued the subpoena with slightly modified language.[18] He was not backing down; the attorney general wanted to know exactly how many of the fifteen thousand New Yorkers hosting on the site fit Airbnb's self-styled picture of benevolent sharing-economy pioneers and how many were simply breaking the 2010 Multiple Dwelling Law to make an illicit buck, taking apartments off the market in the process. In other words, he wanted to know: Did Airbnb hosts in New York City look more like Toshi Chan or Nigel Warren? Were they hoteliers or bed-and-breakfast hosts?

Now Chesky was facing his most difficult decision since the battles with the Samwer brothers and the crisis over EJ. How should Airbnb respond to the subpoena? And what would the cold, hard data reveal to the world about the true nature of its business?

Back in San Francisco, the company was flourishing. That year it had rolled out new mobile apps, refined in the wake of the Snow White exercise, and introduced a feature called instant book that allowed guests to book a selection of Airbnb properties like they would hotel rooms, without time-consuming e-mail exchanges with the hosts.[19]

Those new products contributed to the booming growth. The Airbnb marketplace had the most incredible structural momentum that many of the company's investors and executives had ever seen, driven by a nearly infinite pool of available rooms, apartments, and homes around the world and growing interest among people in a new kind of authentic internet-facilitated travel experience. The company at that point was like a flywheel spinning ever faster, with hosts attracting guests and guests in turn attracting more hosts and an unending stream of headlines about the novelty of the idea accelerating the entire cycle. The problems in New York, meanwhile, were generating a wealth of free, irony-soaked publicity.

Chesky was embracing the moment. His smiling visage appeared on the cover of *Forbes* in January 2013 next to the headline "Who Wants to Be a Billionaire." The young CEO, who only five years before had been full of wrenching self-doubt, was now an adept collector of high-profile mentors, Warren Buffet, Jeff Bezos, and Disney CEO Bob Iger among them. He was determined to think bigger and bolder than any of his colleagues and rivals, according to half a dozen former employees who spoke about this chaotic time in the company's history. Airbnb was on track to earn $250 million in annual revenues in 2013, but Chesky was already thinking ahead to a $2 billion annual run rate. The site had cumulatively booked ten million nights but Chesky was pushing his staff to get to twenty million by the following year. He had around five hundred employees but was projecting a time, only a few years away, when the company would almost certainly have two thousand. Said one senior employee who did not want to be named: "If Airbnb was a body, Nate was the brains, Joe was the heart, and Brian was the balls."

Though Chesky still lived predominantly in Airbnbs around San Francisco, he embraced the glamour that came with this success. In late 2012, he traveled to Asia and Australia with investor Ashton Kutcher and his future wife, actress Mila Kunis. On the trip to Japan to launch Airbnb.com in Japanese, the company ended up spending $15,000 to acquire two matching samurai swords. One former employee says Chesky and Kutcher bought them and later tried unsuccessfully to return them, though the company says that Chesky did not know about the purchase at the time.

In January 2013, Chesky hired a new head of community who shared his devotion to the cause—Douglas Atkin, a former advertising agency executive who had written a 2005 book, *The Culting of Brands: Turn Your Customers into True Believers,* that drew business lessons from devotional sects like the Hare Krishnas. "The opportunity for creating cult brands has never been better," Atkin wrote in the book's epilogue. "Too many marketers have adopted a defensive attitude when actually they are on the brink of creating some of the most tenacious bonds between their brands and customers." Atkin fervently believed that Airbnb wasn't only a company but an ideology and a global movement that existed in a realm beyond provincial laws forged in a dramatically different age.

One of Atkin's first acts at Airbnb was to help start an independent group, called Peers, with the financial backing of Airbnb itself and a mission to support members of the sharing economy. Peers would hold meet-ups in cities where Airbnb and its fellow upstarts faced political hurdles and organize political actions to influence lawmakers. So Atkin's advice to Chesky about the New York battle was clear—he wanted the company to stand up to Eric Schneiderman and fight. Not everyone agreed that it was smart to cross the top law enforcement agent in the state but in the end, Belinda Johnson and Airbnb's other lawyers agreed that the request for data on all of its users was uncomfortably intrusive. "Companies get subpoenaed all the time for information and some companies give it over and some will negotiate it behind the scenes," Johnson later told me. "We made the decision that this was just too broad and we needed to stand up for the privacy of hosts and our community. As public as that was going to be, it was the right thing to do."

And so in October 2013, buoyed by his advisers, the company's relentless growth, a sense of righteousness, and perhaps a feeling of invincibility, Chesky decided to fight the attorney general's subpoena. Instead of handing over the data, Airbnb filed a motion in a New York State court to quash it, arguing that it was unreasonably broad and would violate its customers' privacy. He was essentially telling the attorney general of New York to go pound sand. Observing closely now were lawmakers from cities around the world, including Los Angeles, San Francisco, Barcelona, Amsterdam,

Berlin, Paris, and countless others, who were watching Airbnb spread in *their* cities too. They all had the same worries about home-sharing websites and technologies that appeared to radically disrupt their local economies, with undetermined consequences.

The subpoena and accompanying blizzard of media coverage sent shudders through the ranks of Airbnb users in New York City. Journalist Seth Porges had been renting out a spare bedroom in his two-story duplex in Williamsburg, Brooklyn, since 2010, before Williamsburg was fashionable. At the time his apartment was so geographically inconvenient he had to pitch its proximity to the L Train just so he could fulfill a "bizarre fantasy of being an innkeeper in the countryside and meeting all these amazing characters as they saunter into town."

Two years later, when he was laid off from the men's magazine *Maxim,* his Airbnb earnings let him pursue passion projects instead of taking another full-time job, and he found himself becoming a fervent defender and evangelist for the site. "Airbnb allowed me to be deliberate about my life choices and to think through them and take the risks," he told me. Charging guests around a hundred dollars a night covered his monthly mortgage and more, eventually allowing him to put in another bathroom next to the guest bedroom. Thanks to Airbnb, he was now living in New York City for free.

After the subpoena, Porges, like many other hosts, was suddenly dealing with a torrent of anxiety and misinformation. "I had people book reservations and message me, 'Am I still able to come? What happens now?'" Porges recalls. He offered gentle reassurances. "This is not North Korea." Only using it in a certain way was illegal, and "the police are not knocking down doors. Their beef was with the big illegal hoteliers."

Rich Chalmers, a packaging engineer for a women's apparel company, started using the site after signing a lease with two roommates on a three-bedroom third-floor walk-up apartment on Avenue C, over the Alphabet Lounge, a popular bar. Chalmers found the apartment "so damn loud I couldn't stay in it," so he often went to his girlfriend's place across town. Renting his room on Airbnb to hipster tourists who wanted to stay in Alphabet City started out as a nice source of extra cash.

A year later, while keeping that room, Chalmers also rented a one-bedroom apartment on Ninth Street between First and Second Avenues in the East Village. He then shuttled between the two places and friends' houses, renting both apartments on Airbnb. The rent on the new place was $1,850 a month, which Chalmers easily covered by charging $165 a night and $250 over holidays on Airbnb. "By 2011, I was getting into the swing of things," he told me. He then added a rotating series of friends' and girlfriends' apartments that he would resourcefully list on Airbnb when they were out of town. Suspecting that all this was of dubious legality, Chalmers used an old photograph of himself on the site to reduce the chance that someone might identify him.

As the complexity of his side business mounted, Chalmers asked deli and bodega owners near his apartments to hold spare keys to give to arriving guests. Each place also had a maid who could turn it over in a few hours between guests.

Chalmers estimates that his efforts generated about two hundred thousand dollars in profits over three years, as well as some excellent stories. One day he arrived at his friend Jeff's apartment, which he had rented out on Airbnb, to clean it for a new guest. To his surprise he found the previous guests still there. "They were from Virginia and had come to the city to sell cigarettes and marijuana. I walk in and am like, 'What's going on,' and of course they were so high," he recalls. This would have counted as an Airbnb horror story for some, but not for Chalmers. "I ended up taking them to a different apartment. It was pretty crazy," he says. "The girls were attractive and everyone was up for a party."

After Schneiderman's subpoena, Rich Chalmers, unlike Seth

Porges, thought that it was time to get out. A real estate agent friend told him it was too dangerous and that some landlords were wising up and starting to strictly enforce prohibitions against sublets in their leases. If rent-control laws restricted them from charging market rates for their own properties, they were going to make damn sure that their own tenants weren't going to turn around and reap the full market rate via Airbnb. Chalmers stopped listing in 2012 and paid all the hospitality taxes on his Airbnb income, even erring on the side of caution by refiling for one year.

It was this eclectic mix of earnest hosts and naked opportunists that Chesky was trying to protect when he sent the attorney general back to court to defend his subpoena. "The vast majority of these hosts are everyday New Yorkers who occasionally share the home in which they live," Chesky wrote in an e-mail to hosts on October 7, 2013. "The subpoena is unreasonably broad and we will fight it with everything we've got."[20]

Airbnb commissioned and then released its own survey of its economic impact in New York City, saying it helped to generate $632 million in economic activity for the city in one year, with around 15 percent of that outside of Manhattan.[21] Airbnb visitors stayed an average of six and a half nights and spent nearly $880 at local businesses; by contrast, the average hotel guest stayed four nights and spent $690.

City officials were unimpressed. They were the adjudicators of one of the most difficult choices New York and many other cities had to make, the choice between guaranteeing affordable housing for residents and offering new hotel rooms for out-of-town guests. Airbnb, its critics believed, was removing residential properties from the market as well as deliberately blurring the lines between shared rooms and absentee hosts.

Schneiderman and Airbnb returned to court in April 2014. Airbnb won a temporary victory. A judge ruled the subpoena was too broad because it covered all hosts in the state and not just those in New York City who were breaking the Multiple Dwelling

Law.[22] Schneiderman refiled an amended version a day later, and, its back now pressed against the wall, Airbnb agreed to turn over anonymized data on 16,000 hosts in New York City, including specific information on 124 hosts with multiple listings.[23] The attorney general's office studied the data and five months later issued a critical report that concluded that more than two-thirds of Airbnb rentals in the city violated the law and that a small percentage of hosts with multiple listings were responsible for 37 percent of Airbnb's revenue in the city. Then it created a joint task force with several city departments, including the badly overstretched Office of Special Enforcement, to investigate and shut down illegal hotels in the five boroughs.[24]

In the years following their first meeting in New York, Brian Chesky and Travis Kalanick struck up a sporadic friendship. A few times a year, they would go out to dinner in San Francisco, first by themselves, then with other entrepreneurs or with their girlfriends to discuss their companies' twin successes and their common experiences battling regulators and lawmakers. "I think we learned a lot by watching each other," Chesky says. "There are only so many people in the world that you can relate to [who share] your position."

Employees at both Airbnb and Uber remember these dinners well. Says one Airbnb exec who was also close to Uber employees: "Brian would come back saying, 'We have to be tougher!' and Travis would come back saying, 'We have to be nicer!' "

Executives at Airbnb had watched Uber's travails, following its controversial embrace of ridesharing, and insisted somewhat dubiously that its approach was different and softer than Uber's. "They have their own way of seeking growth," said Jonathan Mildenhall, who joined Airbnb as chief marketing officer in 2014. "I think for us, our community, and the humanity of our community, actually drives a lot of the things we do. So we approach any kind of awkward situation or any challenge with a lot of empathy and a lot of open collaboration... We don't want to kind of bulldoze our way into success. We actually want to partner our way in."[25]

This was consistent with what Belinda Johnson had called the "regulatory brand." But as Airbnb grappled with unfriendly governments in New York and other cities, it turned out that the upstarts were perhaps far more alike than Chesky and his colleagues cared to admit.

Both CEOs talked about their companies with revolutionary fervor. Their handlers now kept each far away from the actual regulatory scrums, worrying that Kalanick might be too combative and Chesky too treacly and earnest. Both were unleashing changes in communities' behavior whose full impact on society they couldn't possibly hope to understand. And each believed that the best tactic was simply to grow, harnessing the political influence of their user base to become too big to regulate.

Chesky's reputation survived this feverish period of empire building far better than Kalanick's. Uber's CEO had discarded political niceties in favor of spirited debate and intellectual sparring, which earned him the image of a pugnacious capitalist. Airbnb's CEO was more circumspect and politically astute, attributes that the freewheeling Kalanick would have to learn in time. But like Kalanick, when confronted by laws that he found unjust, or perhaps just inconvenient, Chesky didn't slow down. His business was every bit as disruptive as Uber's, creating a set of new economic winners and losers.

Reflecting on the years 2011 through 2013, a person might find it difficult to conclude that one company was the more ethical operator. Uber started rampaging over local transportation laws when it appeared competitors might capture strategic ground. Chesky knew that Airbnb violated the strict housing regulations of New York City and elsewhere but pushed ahead anyway, and the site neglected to stop its own users from breaking the law. Both CEOs seized the tremendous opportunities before them with steely determination, pausing just long enough to turn around and repair some of the carnage they left in their wake.

And now it was 2014. Investors outside of Silicon Valley and Wall Street were starting to understand that these startups were special, and they wanted in on the action. Opportunistic entrepreneurs as far away as China noticed, as did European taxi operators, global hotel chains, hotel workers' unions, and all of their powerful allies in government. The upstarts were about to unleash events in Silicon Valley and around the world that neither Travis Kalanick nor Brian Chesky could possibly have envisioned.

PART III

THE TRIAL OF THE UPSTARTS

CHAPTER 10

GOD VIEW

Uber's Rough Ride

Anything you can predict, I expect you to handle.

—Travis Kalanick to Uber CTO Thuan Pham

The Facebook IPO on May 18, 2012, had been a messy affair, with technical problems in the NASDAQ that delayed trading for thirty minutes and a stock price that barely rose at all that first day, then sank into a prolonged slump. The IPO was a litmus test for how the world viewed Silicon Valley and its burgeoning technological revolution. The judgment, it seemed, was harsh.

But over the next year, Facebook executives and their headstrong leader, Mark Zuckerberg, acquitted themselves well. The company restructured its business to exploit the smartphone wave, leading to an uptick in advertising sales and a flourishing stock price four quarters later. On June 31, 2013, Facebook stock exceeded its IPO price, and by the end of that year, it was up a robust 45 percent. That meant that even investors who had piled into the company's later fund-raising rounds, like the Russian investor Yuri Milner, Microsoft, and Goldman Sachs, turned healthy profits.

The triumph of Facebook and its backers would end up changing the course of the upstarts and all of Silicon Valley. At every step of the way, critics had announced that investors were crazy to back the social network at its seemingly overcarbonated valuations. But the conventional wisdom had been wrong. Optimism had paid off handsomely.

Investors tend to ricochet between dueling anxieties: fear of losing money and fear of missing out. Facebook's success suggested that an overabundance of caution in the dawning digital age was misplaced. But it wasn't so easy for pattern-matching investors to simply find and back the next Facebook. Now familiar with the headache of pulling off an IPO and publicly reporting their financials every three months, high tech's premier startups were taking longer to go public. The best, perhaps only, chance for investors was to claw their way into the fund-raising rounds of the hottest private companies.

So financial firms that had historically been tuned to backing public companies started to look for opportunities in private firms, and the flood of new capital into Silicon Valley in the years following Facebook's resurgence spawned heightened competition for deals and pushed valuations up, up, and into the stratosphere. In a span of just six months, the mutual fund manager Fidelity Investments led a round in the image-sharing site Pinterest at a $3.5 billion valuation, while the investment management firm BlackRock led the financing of the online-storage startup Dropbox, valuing it at $10 billion. These were eye-popping numbers Silicon Valley had never seen before in private companies, and they carried the strong whiff of irrationality that had characterized the first dot-com boom. But unlike the situation last time, many of the new internet franchises were popular with consumers and earned real money on advertising and subscriptions. With internet and smartphone penetration growing rapidly around the world, these companies were enticing to users and irresistible to investors.

Uber and Airbnb would emerge from this conflagration of capital and conviction as the twin giants of a new era. By the start of 2014, Airbnb had raised $320 million in venture capital and was valued by investors at $2.5 billion; Uber had raised $310 million and was valued at $3.5 billion. That was trivial compared to what was coming next.

Over the next two and a half years, with Wall Street desperate to capitalize on the success of the upstarts and the Chinese ridesharing giant Didi raising its own enormous war chest to challenge Uber for global supremacy, the two companies together would raise more than $15 billion. They would be worth close to $100 billion before offering a single share of stock to the public.

As the companies swelled in size, value, and ambition, the world grew increasingly concerned about their impact. For Airbnb, the company's influence on housing prices, its effect on residential neighborhoods, and its occasionally awkward attempts at compromise with major cities drew new protests from politicians and regulators. Uber attracted an even larger community of critics by using contract drivers instead of full-time employees, challenging conventional notions of employment, and sparking a seemingly endless litany of controversies related to proper background checks, adequate insurance, and the safety of both the drivers and the riders using its service. Taxi drivers and their representatives, their livelihoods squeezed by Uber and other ridesharing services, led the anti-Uber crusade with angry, sometimes violent protests in countless cities around the world.

The upstarts Uber and Airbnb, frequently named in tandem by sharing-economy proponents and critics, were the defendants in a global trial during this time of uninhibited growth. The issues being litigated were critical: Did the benefits of their dominance outweigh the well-publicized drawbacks? What was their true impact on cities? Were they good for society or bad? Facing these questions, Travis Kalanick and Brian Chesky, both shedding the baggage of their pasts, would have to rise to meet the future with credible testimony on behalf of their companies.

Uber was the first to encounter this towering wall of skepticism. Kalanick and his colleagues were flying high after the introduction of UberX in 2013. Their continued success bred a sense of invincibility and fortified the arrogance that had already permeated their dealings with competitors and regulators. Uber's executives looked down from their perch and, seeing the historic opportunities before them, tried to conquer the world. The world looked up and, for a long moment, wasn't so sure that it liked what it saw.

Back in the summer of 2013, just as Silicon Valley investors were moving from optimism to outright exuberance, Travis Kalanick set out to raise Uber's fourth round of financing. Colleagues say Kalanick set the terms of the financing round himself. He initiated discussions with half a dozen large investors and ran the process as an auction, searching not just for the most capital at the highest valuation, but for a powerful partner who could facilitate Uber's coming global expansion. Yuri Milner's fund, Digital Sky Technologies, was involved in the bidding, as was the venture capital firm General Catalyst Partners. But ultimately Kalanick's attention settled on the dominant technology company in the land—Google.

Kalanick started talks with Google's investment division, Google Capital, but gravitated to its older venture capital group, Google Ventures, or GV, and one of its partners, David Krane. Krane was an early Google PR manager turned investor with a penchant for wearing colorful designer sneakers. He wooed Kalanick with a vision of Google's sixty thousand employees whose collective energies and 20 percent free time at work could be deployed to aid the Uber cause. Kalanick was intrigued by the idea of aligning himself with Google but wanted reassurances from the top and asked for a meeting with founder and CEO Larry Page.

So one evening in August 2013, Kalanick checked into a suite at the Four Seasons Hotel in East Palo Alto, paid for by Google, and woke up the next morning for a ten o'clock meeting with the most powerful man in Silicon Valley. Krane had orchestrated an experience that would blow Kalanick's mind. When Uber's CEO came down to the lobby, a prototype driverless car from the Google X lab idled in front of the hotel, waiting to ferry him to Mountain View. Sitting in the front seat was a Google engineer who could answer all his questions. It was Kalanick's first ride in a self-driving car on real roads.

At the Google campus, Kalanick met with Page, Google senior lawyer David Drummond, and Krane's boss at GV at the time, Bill Maris. Page assured Kalanick that the companies could work together to develop Google Maps, which Uber relied on for navigation in its apps, but he didn't say much or stay very long. The more important legacy that day was Kalanick's developing awareness of the technology that might radically change Uber's business.

"The minute your car becomes real, I can take the dude out of the front seat," Kalanick told Krane excitedly after the meeting. "I call that margin expansion." In Kalanick's estimation, payments to drivers were contra-revenue—a deduction from the top line. The inevitable future of robot cars was going to be awfully good for his business, he surmised.

Krane thought he'd sealed an exclusive investment for Google Ventures after a subsequent four-hour meeting with Kalanick and Uber's head of finance, former Goldman Sachs exec Gautam Gupta. But it wasn't done quite yet. That night, Kalanick called Krane and told him he also wanted to include a second investor in the round: TPG Capital, the San Francisco private equity firm that had engineered leveraged buyouts of such companies as Continental Airlines, J. Crew, and Burger King. Kalanick wanted the experience and connections of TPG's legendary founding partner David Bonderman, then a board member at General Motors, and thought he could help Uber with its regulatory problems around the world.

Google invested $258 million in the ridesharing company. David Drummond joined the Uber board, while Krane joined as a board observer. TPG invested $88 million, buying shares directly from founder Garrett Camp and obtaining a provision that allowed the firm to get additional shares if Uber's valuation ever fell below $2.75

billion, says a person familiar with the deal. Clearly nervous about investing in a startup, the private equity firm was hedging its bets; it also received an option to buy another $88 million worth of stock at the same price within six months. David Bonderman joined the Uber board as a director while his colleague David Trujillo, who had orchestrated the investment, joined as a board observer. (Benchmark also invested another $15 million, and the rapper and entrepreneur Jay Z agreed to invest $2 million—then wired Uber $5 million, hoping for a larger stake. Although Kalanick was impressed by the brash move, he returned the difference.)

Uber's coffers were now brimming. After the round closed, Kalanick climbed aboard TPG's Gulfstream jet with Bonderman, TPG co-founder James Coulter, and Trujillo, as well as investor Shervin Pishevar and his partner Scott Stanford, to visit countries in Asia and gauge the company's expansion opportunities there.

The world seemed wide open. Yet nearly every assumption Kalanick and his investors were making about the future in the fall of 2013 turned out, in the end, to be at least partially incorrect. Google was reluctant to cede the results of its driverless car research to another company and would soon look more like Uber's mortal enemy, not its ally. Within a year, David Bonderman would leave the board of General Motors, which in 2016 would make a sizable investment in archrival Lyft.

And remarkably, according to multiple people familiar with the transaction, when the time came for TPG to purchase its second $88 million allotment of Uber shares at the same valuation, the private equity firm wavered and waited until the last possible moment before attempting to exercise the option. Characteristically stingy about giving out Uber stock and diluting the ownership stakes of existing investors, Kalanick declined the transaction. Calculating for the dramatic rise in Uber's value between that round and the end of 2016, TPG's lack of faith ended up costing the firm hundreds of millions in unrealized gains.

The biggest miscalculations may have been Kalanick's own.

Asia would prove more challenging and costly than he had ever anticipated. He especially misread the atmospheric shifts in Silicon Valley's fund-raising climate. "Emil," he had said gleefully to Emil Michael, his new vice president of business development, after closing the investment from Google and TPG, "we're never going to have to fund-raise again."

Emil Michael was disappointed to learn that Kalanick thought Uber's financing efforts were over—he considered fund-raising one of his talents. Born in Cairo, Michael had immigrated with his family to the United States as an infant, graduated from high school in New Rochelle, New York, and earned an undergraduate degree from Harvard University and a law degree from Stanford. He had a brief stint at Goldman Sachs before decamping to Silicon Valley in 1999, right at the peak of the dot-com bubble.

During his ten years in the industry, Michael had cultivated a reputation as being effective, loyal, and upbeat. He first met Kalanick in 2011, when he was taking a hiatus from high tech to work in the White House as a special assistant to Secretary of Defense Robert Gates. Kalanick tried to recruit him to join the startup, but at the time Uber looked like a luxury town-car service, not a worldwide transportation juggernaut. Michael was skeptical that it could ever be a big business.

But Michael remained friendly with Kalanick and by the time he joined Uber, in the fall of 2013, he recognized that Uber's future was brighter than he had originally believed. While Uber Black remained one and a half times more expensive than a traditional yellow taxi, UberX was, on average, 25 percent *less* expensive and was starting to dominate the emerging rideshare wars.

Lyft and Sidecar had introduced ridesharing, but when Uber started aggressively rolling out the service, first in the United States in 2013 and then in Europe in 2014, the two rivals struggled to keep up. Uber had a more established brand and more money in the bank as well as upscale product lines, like Uber Black and Uber SUV, whose profits could be used to subsidize UberX rides and offer financial incentives to new drivers.

Uber was growing 20 percent each month and, thanks to UberX, had gone from nonexistent to ubiquitous nearly overnight in San Francisco, Los Angeles, DC, and Boston. That fall, Uber had moved out of its cramped offices on Howard Street to more spacious digs a few blocks away, on the ninth floor of 706 Mission Street, around the corner from the San Francisco Museum of Modern Art. Kalanick's desk was across from Emil Michael's, and the two would often peer at each other over their computer screens to marvel at new growth statistics.

"We'd have these moments, asking each other, 'Did you see this thing?'" Michael says. "It just kept going."

Some U.S. cities, such as Austin, Las Vegas, Denver, and Miami, resisted the arrival of unregulated ridesharing; amusingly, New Orleans sent Uber a cease-and-desist letter before it was even operating there.[1] But Kalanick still had his trusty playbook as well as the political theorem known as Travis's Law, which dictated that politicians accountable to the people could be pressured to accommodate any service that was markedly better than the alternative.

In October 2013, most of Uber's four hundred employees flew to Miami on another workation, staying in rooms in the ritzy Shore Club in South Beach. When employees weren't at dinners or parties around the hotel pool, which had the giant *U* from the Uber logo illuminated on the water, they walked the beach handing out Uber postcards and affixing pro-Uber posters to light poles. The company's campaign to drum up popular support to legalize ridesharing in South Florida had a website, an Instagram page, and a Twitter hash tag: #MiamiNeedsUber.

Miami was a challenging market for Uber. Private for-hire limos and sedans were required by law to wait an hour before picking up passengers and had to charge more than seventy dollars for the ride. The ordinance was backed by the region's taxi fleets and meant to keep them safe from loosely regulated competition from limos and town cars. It didn't stand a chance against sustained popular demand for ridesharing. Lyft and then Uber would open for business in Miami-Dade a few months after the visit by Uber employees.[2] Though the companies' services were still technically illegal, courts only occasionally levied fines against drivers, and the police didn't shut down either service. By 2015, lawmakers were ready to change the rules.

"Demand is too great," Miami mayor Carlos Gimenez told the *Miami Herald*. "I'm not going to drag Uber and Lyft back into the 20th century. I think the taxi industry has to move into the 21st."[3]

Uber was entering adolescence, winning political battles, growing, and adding executive talent. A few months before Emil Michael joined the company, Kalanick had also recruited a new chief technology officer, Thuan Pham.

Pham had left Vietnam as a child, spent ten months in an Indonesian refugee camp, attended MIT, and became an accomplished technical leader at the online advertising firm DoubleClick and cloud company VMWare. Joining Uber as a senior executive meant a grueling interview process that included a cumulative thirty hours of one-on-one conversations with Kalanick. Pham reorganized Uber's technical team, accelerated the hiring of engineers, and oversaw a complete revision of its dispatch algorithms and database storage systems to keep up with a business that was doubling every six months and showing no signs of slowing down.

Pham's impact at Uber was evident on New Year's Eve, typically a night of frantic activity that had overwhelmed Uber's systems for three straight years. "Thuan, if we have a system breakdown, I'm going to have an aneurysm and my death will be on your hands," Kalanick told him earlier that day. But for the first time, Uber's systems survived the night relatively unscathed. A few days later, Kalanick took Pham and his team out for a celebratory dinner and offered a rare bit of praise. "You did a great job," Uber's CEO said. Characteristically, the praise came with a new challenge. "From here on out, anything you can predict, I expect you to handle."

Over the next few months, Kalanick executed two ideas that further propelled the growth of UberX. The first, helping Uber drivers finance the lease of new vehicles, originated with former Goldman Sachs commodities trader Andrew Chapin, who was working as a driver operations manager in Uber's New York office. Chapin had observed that the biggest obstacle facing many prospective Uber drivers was the lack of a vehicle; a lot of them didn't own cars because they were immigrants with bad credit or no credit.[4]

Chapin thought Uber could help drivers obtain car leases and then divert a certain percentage of their earnings toward paying them off. The arrangement would pay dividends for the company, not only by putting more cars on the road but by ensuring that drivers devoted their energies to Uber rather than to rival ridesharing or delivery services. "The demand is there, but if we don't help our partners and drivers get cars on the road, then it just doesn't matter. We're just not going to be able to grow," Kalanick said that year.[5]

To canvass for interest in such a program, Uber executives visited car companies and auto-loan financers around the country. Their initial reaction was skepticism. "The car companies were like, 'Yoober? Who are you guys? Aren't you the town-car company?' " Emil Michael recalls. Kalanick, Michael, and investor Bill Gurley visited the Detroit offices of Ford Motor Company, often referred to as "The Glass House," and met with executive chairman William Clay Ford Jr., who was also noncommittal. Kalanick got a photo of himself with Ford, the great-grandson of founder Henry Ford, plus a tour of the company's historic displays in the lobby, where Gurley recalls the Uber CEO got lost in reading about the automaker's storied past.

Ultimately, the big carmakers, GM, Toyota, and Ford, would sign on to the program, as would dealerships and auto lenders, and in time Uber would bring the financing in-house and make loans through its own subsidiary, Xchange. The program would be criticized for offering subprime loans with onerous terms and for repossessing vehicles when drivers didn't make their payments on time.[6] Michael argued that the program helped credit-challenged drivers who simply had no other options. "You are taking people who are already getting killed on loans and doing something better for them," he says. "Of course the interest rate is high, but at least they have a chance."

While driver loans helped stimulate the supply of Uber cars, a second move helped to spark demand and was just as controversial. In early 2014, hoping to improve business during the annual winter slowdown, when people curtailed their nights out, Kalanick cut UberX fares by up to 30 percent in U.S. markets like Atlanta, Baltimore, Chicago, and Seattle.[7] The theory was that if prices went down, customers would use the service more and bypass rental cars, public buses, and subways. With more passengers, drivers would spend less time idling between rides, replacing the lost income from the fare cuts by completing more rides.

While the plan made sense, subsequent fare cuts would create unrest among drivers, and Uber would eventually have to reverse them in cities where lower prices didn't spark higher demand. But it also accelerated the growth of UberX and, perhaps just as important, forced the less highly capitalized Lyft to introduce its own fare and commission cuts.[8] Uber had discovered what startup gurus like to call the virtuous circle, the links between various parts of its business. Lower prices led to more customers and more frequent usage, which led to a larger supply of cars and busier drivers, which enabled Uber to further cut prices and put more pressure on competitors.

Even Uber's most fervent supporters had not grasped the true potential of the business. Uber wasn't just taking passengers out of yellow cabs, it was growing the overall market for paid transportation.

"I knew Uber was going to be big, but I didn't know it was going to be so outlandish," says venture capitalist Bill Gurley. "When we started testing lower price points, that's when it was really 'Oh my God.' The price elasticity was impressive." The increasing pace of the business surprised Kalanick himself. "I didn't understand the scope of the Uber opportunity and I didn't understand how the private equity and venture worlds would go to massively unprecedented places in order to be a part of that opportunity," he says.

Nothing could stop Uber now, it seemed, except perhaps itself.

On December 31, 2013, just before eight o'clock on New Year's Eve, a young mother named Huan Kuang and her two children were walking in a Polk Street crosswalk in San Francisco's Tenderloin District when tragedy struck. A gray Honda Pilot SUV turned right onto Polk, hit the family, and killed Kuang's six-year-old daughter, Sophia Liu. The driver of the vehicle, fifty-seven-year-old Syed Muzaffar, had been working for UberX for about a month. He didn't have a passenger in his car but told police he was monitoring the Uber app, waiting to get assigned a fare. The distraught mother later told a local television reporter that she could see the light of the cell phone reflected on the driver's face.[9]

The media reeled at the circumstances surrounding the case and at the uncomfortable fact that, at first, Uber denied any involvement. *We can confirm that this accident DID NOT involve a vehicle or provider doing a trip on the Uber system,* Travis Kalanick Tweeted the afternoon after the accident.[10] When more facts emerged, Uber released a more carefully worded abrogation of responsibility that, even coming after a statement of condolence to the family, reeked of cold, calculating legal logic. "The driver in question was not providing services on the Uber system during the time of the accident," read a statement on the Uber blog posted the day after the incident. "The driver was a partner of Uber and his account was immediately deactivated."[11]

Uber's stance defied any common-sense assessment of the tragedy. The company was earning significant money but apparently wanted none of the liability for accidents. Yet Uber had brought drivers like Muzaffar onto the roads with the lure of a profitable night, furnished them with a smartphone app, and deployed a system that required them to respond promptly to multiple alerts and text messages while on the road. Muzzafar might not have had a passenger in the backseat, but he was performing a key service for Uber: driving around the city with the app open waiting for one.

There were other disturbing facts. Muzaffar had a ten-year-old citation for driving a hundred miles an hour on a highway in the Florida Keys.[12] Uber's background checks were done by a company called Hirease, which reported only the previous seven years of an individual's criminal history.[13] Uber carried insurance with $1 million in liability coverage per incident but that applied only from the moment a driver accepted a trip on the Uber app to when the passenger left the car. Neither Uber nor the California Public Utilities Commission, in its contentious hearings the year before, had considered the period when drivers were logged on to the system but had empty backseats and were waiting for fares. Sophia Liu's family, which had significant medical bills, would have to rely on Muzaffar's terribly inadequate $15,000 personal insurance policy, if it even paid up. Such a tragedy had been eminently predictable, and yet Uber, it seemed, hadn't been ready for it.

(That March, three months later, both Uber and Lyft introduced up to $100,000 of supplementary insurance to cover this gap.[14] In 2014, the State of California passed a law mandating the companies carry $200,000 worth of liability coverage for the period when drivers had the app open and were looking for passengers.[15])

Muzaffar was arrested on vehicular manslaughter charges after the incident and tried. The jury deadlocked in April 2016,[16] and, as of this writing, he is awaiting retrial. The Liu family also filed a wrongful death lawsuit against Uber, alleging the company was responsible because its smartphone application had fatally distracted Muzaffar from the road. Uber settled the suit privately in June 2015 for an undisclosed amount while admitting no wrongdoing.[17] Yet the damage to Uber's reputation went far beyond any secret payment. For the first time, the company was widely seen as either unable or unwilling to contain the potentially destructive consequences of the transportation revolution it was unleashing on the world.

Uber execs "were extremely hungry and immature and caught up in a whirlwind of money and growth," says Christopher Dolan, a local plaintiff attorney who represented Sophia Liu's family. "They got seduced by the possibility rather than stopping to think about their responsibility."

The Sophia Liu tragedy kicked off a year of relentlessly negative media coverage that calcified Uber's reputation as an aggressive, ruthless, and sometimes heartless operator. While Uber spread rapidly into cities and countries in Europe and Asia, critics slammed it not only for facilitating dangerous conduct by drivers with backgrounds that hadn't been thoroughly vetted but also for what appeared to be anticompetitive tactics and the occasional inappropriate public comments by employees. Many people's negative impressions of Uber were forged in 2014, during the company's most challenging year, when mistakes by Kalanick and his team seemed to compound as their business grew.

Less than a month after the Liu tragedy, an unseemly practice among the ridesharing startups roared into public view. Uber, Lyft, Sidecar, and a host of smaller players were only as strong as the number of drivers willing to open their apps, so they constantly vied not only to capture new drivers but to poach one another's. It was bare-knuckle competition, the kind of junkyard brawling that Travis Kalanick loved—so Uber excelled at it. Inside the company, employees called it slogging, only later ginning that word into an acronym for a meaningless term: *supplying long-term operations growth*.[18]

Mostly it involved offering free gas cards, signing bonuses, and other perks to get drivers from rival services to defect, but occasionally it went further. On January 24, 2014, the Israeli startup Gett reported that over a three-day period in New York City, where it had introduced a black-car service, Uber employees ordered and then canceled more than a hundred of its cars and then texted the drivers and attempted to get them to switch to Uber. (Gett, unlike other ridesharing companies, foolishly was not using a service such as Twilio to disguise driver phone numbers.) Jing Wang Herman, Gett's U.S. manager, compared Uber's strategy to an attack by hackers and sent Gett drivers a text message apologizing for the harassment and declaring, in all capital letters, WE ARE AT WAR WITH UBER. She also showed media outlets a list of Uber employees who had ordered Gett cars using their real names. It included Josh Mohrer, the general manager of Uber's New York office.[19]

Confronted with the evidence, Uber promptly apologized. "The sales tactics were too aggressive," read a post on the company's blog. Kalanick later told me, "The New York team works hard to get as many drivers onto the system as possible, because that's the only way to grow and service customers in the city with quality, reliability, and the right price. Sometimes they get a bit aggressive, and that's unfortunate. We apologized and made it a lesson for the rest of the company."

Uber's year would become only more contentious from there. In February, a feature in the men's magazine *GQ* described Kalanick as a "bro-y alpha nerd" and quoted him observing that Uber had amplified his appeal to women. "Yeah, we call that Boob-er," he said.[20]

In May, speaking at the Code Conference, Kalanick had even more trouble elevating his tone to a level befitting a high-profile CEO. I was in the audience that year when he attacked incumbent cab companies so forcefully that it made the taxi fleets look sympathetic. Uber, he said, was engaged in a political campaign where "the candidate is Uber and the opponent is an asshole named Taxi. Nobody likes him, he's not a nice character, but he's so woven into the political machinery and fabric that a lot of people owe him favors." He went on to say Uber would have to "bring out the truth about how dark and dangerous and evil the taxi side is."

When asked about driverless cars, he said that he was excited for the technology because it could bring prices down, but he didn't express concern about unemployment for drivers. "The reason Uber could be expensive is because you're not just paying for the car, you're paying for the other dude in the car," Kalanick said. As for the tens of thousands of drivers who relied on his company to support their families, he shrugged. "This is the way of the world," he said, "and the world isn't always great. We all have to find ways to change."

Kalanick was being himself—blunt and unaware of, or perhaps indifferent to, how his comments might be perceived by Uber's own key constituencies. Uber's problems in 2014 were reflections of its CEO's personality, the strengths that had gotten him through the ordeals of his early career and the flaws that had at times repelled some investors and colleagues. Chief among those was his furious competitive streak, his drive to win not only in Wii Tennis but in business and to rub out his rivals in the process.

The battle with Lyft, which continued to play out in 2014, was another example of this. Kalanick had regretted letting Lyft take hold in 2012 while waiting for California regulators to sanction ridesharing. He was obsessed with Lyft and its potential to outmaneuver Uber, and he worried that a more seasoned company might acquire it.

Around this time, he confronted his fellow upstart CEO Brian Chesky while he was dining at The Battery, a swank, members-only social club for the San Francisco tech set. Chesky was having drinks with local attorney Sam Angus when Kalanick came up to their table and demanded to know if Airbnb was going to buy Lyft.

"No, we are in the business of trips," Chesky recalls answering.

"*We* are in the business of trips!" retorted Kalanick, who later couldn't recall whether he was joking or if he had been responding to an actual rumor.

For a brief period in 2014, Lyft had been ready to throw in the towel, and representatives approached Uber about combining the companies. Kalanick and Emil Michael went to dinner with Lyft president John Zimmer and Andreessen Horowitz partner John O'Farrell to discuss a deal, according to three people who were privy to the conversations. The meal was friendly, despite the heated rivalry. But Lyft's expectations were high. In exchange for selling Lyft to Uber, Lyft's backers wanted an 18 percent stake in Uber. Uber offered 8 percent; Kalanick wasn't a fan of mergers to begin with and wasn't about to hand over a fifth of his prize. Neither party would budge, and the talks fell apart.

Lyft recovered quickly. That spring, with unconventional sources of capital now flooding into Silicon Valley, it raised $250 million from a consortium of investors that included hedge fund Coatue Management, Chinese e-commerce giant Alibaba, and the Founders Fund, the investment vehicle of PayPal co-founder Peter Thiel, and it expanded into twenty-four new U.S. cities, thirteen of which were midsize markets where Uber did not yet operate.[21]

The battle was on again. A few weeks later, Uber rushed to raise another $1.2 billion in a hastily convened Series D round from the financial firms Fidelity, Wellington, and BlackRock, as well as the venture capital firm Kleiner Perkins. The fund-raising process took all of three weeks, and Kalanick was at his charismatic best, pitching investors a compelling vision of Uber's future.

"If you can make it economical for people to get out of their cars or sell their cars and turn transportation into a service, it's a pretty big deal," he told me after the round closed. Kalanick also took the unusual step of privately telling investors that if they wanted a chance to make an investment in Uber, they shouldn't consider talking to Lyft.[22] Uber was "going to Lyft's investors and saying,

'Hey, look, we're open for business too. We'll take an investment,'" Kalanick said when I asked him about the tactic. "That's what that conversation was." But to others, it seemed Kalanick was trying to scorch the earth behind him.

It must have seemed unfair to Kalanick that Lyft had the better reputation, even though in some ways it was the more aggressive player. It had been the first to introduce unregulated ridesharing in San Francisco, Miami, and Kansas City, yet the endeavors of its founders, Logan Green and John Zimmer, often came off as sincere idealism, not predatory ambition. "Every Lyft ride is an opportunity for positive human interaction, Zimmer gushed to CNN in one characteristic interview. "I also feel very fortunate to be changing the future of transportation, which will deliver a more people-centered city of tomorrow."[23]

That July, Lyft started preparing to launch ridesharing in New York City, where Uber operated only with licensed professional drivers. Sidecar had attempted such a feat the year before, only to see its drivers issued summons and their cars impounded by the New York Taxi and Limousine Commission. It beat a hasty retreat.[24] As Uber had discovered during the introduction of Uber Black and Uber Taxi in New York, the TLC could be a formidable adversary that did not tolerate disruption in the city's already jammed streets.

But Lyft president John Zimmer wouldn't take no for an answer. He announced publicly that Lyft would introduce its service in Queens and Brooklyn.[25] Then Zimmer flew to New York City with his vice president of government relations, David Estrada, and had a day of meetings with Meera Joshi, the taxi commissioner under New York mayor Bill de Blasio. Joshi informed them in strident terms that Lyft needed to register as a base and, like Uber, could use only TLC-licensed drivers. The next day, Zimmer and Estrada were called to the state attorney general's office, where a dozen officials from the AG's office and New York's Department of Financial Services reeled off a list of laws Lyft would be breaking if it followed through with its plans.

Still determined, Zimmer hosted a launch party that night at the 1896 nightclub in Bushwick, with a performance by the rapper Q-Tip from a Tribe Called Quest. Local techies crowded the dance floor while a dozen taxi drivers protested out front. "We feel Lyft is coming in here to take us out of business," Nancy Soria of the New York Association of Independent Taxi Drivers told the tech blog *Technical.ly.*[26]

Later that night, Zimmer and Estrada heard the TLC was preparing an injunction. On a conference call with general counsel Kristin Sverchek and Lyft's outside lawyer, an impassioned Zimmer argued that they should go ahead anyway and wanted to get himself arrested for the cause. The lawyers laughed—but he was serious. Together they persuaded him that it would be a bad idea. "I don't want to think about you in jail," Sverchek told him. "It's not something I can stomach."

Backed into a corner, Lyft caved. For the first time in its history, Lyft introduced a service using professional drivers instead of regular people driving their own cars.[27] In New York, Lyft would look like the original incarnation of Uber, deploying only licensed drivers.

From there the battle between Uber and Lyft devolved further. In public, they accused each other of slogging—ordering and canceling rides and proffering rewards to each other's drivers to defect.[28] In private, an even more rancorous struggle played out. Lyft's chief operating officer was a creative executive in his early thirties named Travis VanderZanden whose fledgling on-demand car-wash business, Cherry, Lyft had acquired in 2013. At Cherry, VanderZanden had devised an ingenious system in which the most experienced car washers mentored and reviewed newer ones, allowing the company to enlist a large workforce of contractors without having to hire employees to train and monitor them.[29] VanderZanden brought the idea to Lyft and used it to help expand Lyft into new cities without ever putting employees on the ground. And he introduced an Uber Black–like service, Lyft Plus, an attempt to cut into a source of Uber's profit advantage.

But by the summer of 2014, his colleagues from that time said, VanderZanden had become disillusioned about Lyft's prospects against the more highly capitalized and faster-moving Uber. Without the knowledge of Green or Zimmer, he approached two Lyft board members about taking over as CEO, according to court filings.[30] He also started talking privately to Uber about restarting merger discussions between Lyft and Uber. When Lyft's founders found out about all this, they were livid. VanderZanden resigned in August and a few weeks later joined Uber as vice president of international growth.

The lawsuits promptly flew. Lyft accused VanderZanden in California state court of downloading proprietary financial and strategic documents before he left.[31] VanderZanden denied the allegations and on Twitter called it an "audacious attack on my reputation."[32] A few months later, Uber filed a civil lawsuit in federal court in an attempt to find out who had illegally breached its computer systems and downloaded the names and personal information of some fifty thousand drivers. According to a deposition in VanderZanden's case, Uber believed the culprit was Lyft's chief technology officer Chris Lambert. (Lambert's lawyer denied to Reuters that the Lyft CTO had anything to do with the data breach.)[33]

Things had gotten ugly fast and the public airing of all this enmity was going to be bad for business. So, two years later, right before the VanderZanden case was set for a potentially embarrassing public trial, the two companies settled their legal dispute and Uber withdrew its civil lawsuit over the data breach.[34] The heated rivalry would continue to play out, but on the streets, not in the courtroom, and not on Twitter.

* * *

That summer, the rapidly growing Uber moved its headquarters for the second time in a year, and the seventh time since its founding, to larger offices, taking over eighty-eight thousand square feet in a former Bank of America building on Market Street and leasing extra space for expansion. The hulking cement structure occupied an entire city block and had a helipad on the roof and a bank vault in the basement. Inside, Uber's offices were dim and moody, full of dark wood, chocolate-brown leather sofas, and walls covered by whiteboards and digital displays of Uber's cities. A circuitous walking path wound amid the open desks, perfect for Kalanick's restless pacing. True to the CEO's combative personality, the long executive boardroom at the center of the main floor, with clear glass walls that could be frosted opaque during private discussions, was dubbed "the war room."[35]

While the name suggested deeply ingrained belligerence, Uber that summer was desperately trying to professionalize its image. In August, after a protracted process of interviewing political luminaries like Democratic strategist Howard Wolfson and White House press secretary Jay Carney, Uber made a high-profile hire—David Plouffe, the manager of Barack Obama's 2008 presidential campaign, became senior vice president of policy and strategy.[36] Kalanick also set about tempering his tone, and he introduced a more inspirational articulation of the company's mission; it was no longer to destroy "an asshole named Taxi" but to offer "transportation as reliable as running water, everywhere and for everyone."[37]

Kalanick was trying to change his ways. But it would take more than a political craftsman to reshape Uber's pugnacious and well-earned image. Uber's biggest public relations crisis of the year was still to come.

In late October, journalist Sarah Lacy of the technology blog *PandoDaily* wrote an essay excoriating Uber, in part for a ridiculous promotional campaign out of its office in Lyon, France, that offered to pair riders with attractive female drivers who appeared to be affiliated with an escort service.[38] "Who said women don't know how to drive?" read Uber's advertisement; around the post were pictures of scantily clad women.[39] When contacted by the press about the promotion, Uber quickly canceled it and pulled the post from its local Lyon blog. But Lacy, never one to shy away from bombastic pronouncements, declared that she was deleting the Uber app from her phone and accused Uber of having a corporate culture of sexism that endangered female drivers and passengers.

"I don't know how many more signals we need that the company simply doesn't respect us or prioritize our safety," she wrote.[40]

Inside Uber, Lacy's post went over badly. The Lyon promotion had been an embarrassing mistake by a local office but the company prided itself on extending opportunities to women to become drivers and ensuring that, as passengers, women could order cars safely and not have to wait on dark street corners at night hoping to hail a passing cab. Lacy's post, on top of all the mounting criticism that year, rankled.

Three weeks later, Uber invited various media executives and journalists to an off-the-record dinner at Manhattan's Waverly Inn. Kalanick sat on one side of the long table and, after dinner, gave a short speech and answered questions. Emil Michael sat on the other side, across from *New York Daily News* publisher Mort Zuckerman and Arianna Huffington. Next to them sat Ben Smith, the editor in chief of *BuzzFeed,* who asked Kalanick in the Q-and-A session how he felt about the Affordable Care Act, aka Obamacare. When the table returned to private conversations, Michael asked Smith why he had posed a political question, and Smith volunteered that he was hoping that Kalanick would give an answer that revealed libertarian leanings.

That led to a broader conversation about the news media and its scruples, the exact content of which would become a topic of blistering controversy. Michael's recollection of the discussion is that he told Smith that it bothered him when the press made personal accusations without evidence. Then he raised the hypothetical idea of Uber spending a million dollars to create a coalition for responsible journalism; he said it could hire researchers and professional journalists to respond when negative articles came out and turn the tables on reporters, who might have their own secrets to hide.

The dinner was on a Friday. On Monday, claiming he didn't know the discussions at the event were off the record, Smith published an article on *BuzzFeed* with his version of the conversation, headlined "Uber Executive Suggests Digging Up Dirt on Journalists." Uber's muckrakers would "look into 'your personal lives, your families,' and give the media a taste of its own medicine," Michael had said, according to Smith. He also reported that Michael theorized that the reporters could unearth unsavory details about Sarah Lacy's private life and said that she should be held "personally responsible" for "any woman who followed her lead in deleting Uber and was then sexually assaulted."[41]

Though Michael privately disagreed with Smith's characterization of his words, he immediately apologized in a statement, saying, "The remarks attributed to me at a private dinner...do not reflect my actual views and have no relation to the company's views or approach. They were wrong no matter the circumstance and I regret them." A day later, in an account of the same conversation in the *Huffington Post,* Nicole Campbell, a White House fellow who was sitting nearby, framed Michael's comments somewhat differently. Campbell wrote that Michael had said hypothetically that Lacy "wouldn't like it if someone wrote false things about her or published an article that was factually wrong, because we all have done things in our private lives we are not proud of."[42]

The *BuzzFeed* story contained one additional element that fueled the immediate backlash against Uber. Smith wrote that a few days before the dinner, the general manager of the New York office, Josh Mohrer (he of the Gett slogging incident from earlier in the year), had greeted visiting *BuzzFeed* reporter Johana Bhuiyan outside the Uber office in Long Island City and showed her that he had been tracking her journey in an Uber car using a company tool called God View.

God View was an internal service that Uber made available to all of its employees, and it was one reason the company had grown so quickly. All of the hundreds of city offices had access to the same tools as employees in San Francisco and so could make decisions based on data in a decentralized way. Kalanick believed that a transparent corporate culture gave employees a sense of ownership of their projects, making them feel as if they were running startups within the larger company. And yet Uber had deployed God View without adequate privacy protections around the data, with little training of employees, and with no public privacy policy that informed the world of how Uber planned to use this sensitive information. It was a disaster waiting to happen.

The *BuzzFeed* story, with its implications of menacing corporate behavior and misused customer data, set off a media bomb. After a year of drama, news outlets were attuned to any whiff of Uber-related controversy, so nearly every major publication and television network covered the story. It was picked up as far away as Europe and Asia. The next morning, Michael and Kalanick left New York City for a Goldman Sachs conference in Las Vegas. Michael recalls walking the concourse at LaGuardia Airport with Kalanick and glancing up at a television in an airport lounge to see his picture on CNN.

It all seemed surreal. On the plane, Michael and Kalanick sat side by side with their laptops connected to the in-flight Wi-Fi and watched as a torrent of anti-Uber Tweets rolled in reacting to Michael's comments at the dinner. "I was literally trying to distract him," Michael recalls. "I was thinking, *Oh my God, I'm going to get fired before we land.*" He had never blundered in such a public way before.

At a previous point in his career, Kalanick might have gone to war with his online critics, defensively seeking to protect his beloved brand. Instead, he took to Twitter and, with Michael sitting next to him trying not to look at his boss's laptop screen, reeled off fourteen Tweets that temporarily quelled the storm. Then he laid out a promise that Uber would strive to be a better citizen of the world.

> Emil's comments at the recent dinner party were terrible and do not represent the company.
>
> His remarks showed a lack of leadership, a lack of humanity, and a departure from our values and ideals.
>
> His duties here at Uber do not involve communications strategy or plans and are not representative in any way of the company approach.
>
> Instead, we should lead by inspiring our riders, our drivers and the public at large.
>
> We should tell the stories of progress and appeal to people's hearts and minds.
>
> We must be open and vulnerable enough to show people the positive principles that are the core of Uber's culture.
>
> We must tell the stories of progress Uber has brought to cities and show our constituents that we are principled and mean well.
>
> The burden is on us to show that, and until Emil's comments we felt we were making positive steps along those lines.
>
> But I will personally commit to our riders, partners and the public that we are up to the challenge.
>
> We are up to the challenge to show that Uber is and will continue to be a positive member of the community.
>
> And furthermore, I will do everything in my power towards the goal of earning that trust.
>
> I believe that folks who make mistakes can learn from them — myself included.
>
> And that also goes for Emil.
>
> And last, I want to apologize to Sarah Lacy.

In Las Vegas, Michael stayed in his hotel room, far from the conference at the Bellagio. Back at the office later in the week, Kalanick gathered his employees, many of whom were distraught by the public backlash, and addressed the entire mess, saying he trusted Michael and was certain that the senior executive misspoke and did not have nefarious intent. He was not going to fire him.

But Kalanick also conceded the company was so big, powerful, and vital to urban transportation that it had to grow up. The gunslinger mentality that was such an asset during Uber's first few years was now causing more harm than good. Access to God View had to be strictly limited and controlled if the company wanted to preserve the trust of its users.

Even he, as CEO of the most closely watched startup in the world, had to change his tone, become more self-aware, and articulate the future Uber was rapidly creating with optimism and a whole lot more empathy.

A few days before Kalanick had spoken at the Code Conference and declared war on "an asshole named Taxi," he had received a phone call from Google chief lawyer David Drummond, an Uber board member. Google co-founder Sergey Brin was also speaking at the event and, Drummond told him, was going to make a bombshell announcement: Google was planning to roll out its driverless cars as part of its own Uber-like on-demand service. Drummond wanted to give Kalanick advance warning that the search giant was going to unveil long-term plans to compete with Uber.

An hour after the call, Drummond phoned again, saying the announcement was off—Brin wasn't going to say that after all. Kalanick was stunned. In many ways Google was an impressive, well-run company, but as he was learning, inconsistencies often resulted from the impetuous whims of its founders.

Nevertheless, the experience planted an unsettling thought in Kalanick's head. The company he had considered an investor and ally just eight months ago now loomed as possible competition. High-tech history was replete with examples of technology companies whose prospects were curtailed by dependencies—IBM's reliance on Microsoft for the Windows operating system in the 1980s, for example, and Yahoo's dependence on the Google search engine in the 2000s. Uber required Google for maps, and perhaps one day it might need Google even more for its driverless cars.

That fall, as the public relations crises mounted, Kalanick was secretly preparing for this contentious future. He started meeting regularly with his new chief product officer, Jeff Holden, a fast-talking former Amazon and Groupon executive who had spearheaded the launch of the carpooling service Uber Pool, and with Matt Sweeney, an early engineer who had orchestrated a complete overhaul of the Uber application. In October, Kalanick received additional confirmation from inside Google that the search giant was planning to compete with Uber. He subsequently asked board member David Drummond and board observer David Krane to stop attending Uber board meetings.

Kalanick and his executives were plotting how they could jump-start Uber's own driverless-car project and catch up to Google and electric-car maker Tesla. If the future of transportation was indeed going to be driverless, they reasoned, Uber had to own it.

CHAPTER 11

ESCAPE VELOCITY

Fights and Fables with Airbnb

Victor Hugo had a saying: You cannot kill an idea whose time has come. And our time has come.

—Brian Chesky[1]

A few months before Uber moved to its mood-lit offices on Market Street, Airbnb left its own comfortable nest, exchanging its Potrero Hill space for swank new headquarters a five-minute walk away in a hundred-year-old warehouse on 888 Brannon Street. Brian Chesky and Joe Gebbia unleashed their design talents on a building that had served at various times over the years as a wholesale jewelry market and a battery factory. They installed a light-dappled atrium with a twelve-hundred-square-foot vertical "green wall," made up of hundreds of plants spanning three floors, and outfitted a dozen conference rooms to look like Airbnb listings in Milan, Paris, Denmark, and elsewhere. Another conference room was modeled after the founder's original apartment on Rausch Street, while a larger meeting space was converted into an exact replica of the war room in the Stanley Kubrick film *Dr. Strangelove,* complete with a circular table under a ring of Cold War–era overhead lighting.

No expense was spared in Airbnb's new office. There were pricey aluminum Emeco chairs, gold-plated tableware from a local ceramics boutique, and a gourmet kitchen that served three meals a day, seven days a week. The company says it spent more than $50 million on the renovation and $110 million on a ten-year lease (a deal that would look good as city rents soared).

At a board meeting, according to a person who was there, venture capitalist Marc Andreessen expressed concern over the company's exorbitant burn rate. Another board member, Sequoia Capital's Alfred Lin, who had replaced his former partner Greg McAdoo on the Airbnb board, confirmed there were discussions about profligate spending but said they were overshadowed by the company's remarkable performance. "Growth covers a lot of sins, and the growth of the company was spectacular," Lin says.

The new headquarters wasn't designed merely as an office but also as a shrine to an idea—that Airbnb could bring people together, erase their differences, and, in the earnest spirit of so many Silicon Valley satires, *make the world a better place.* On the third floor, near the visitor check-in desk, a street sign gifted to the company by the director Spike Lee identified the foyer as "Do the Right Thing Way." The walls of the office were covered with inspirational phrases, such as AIRBNB LOVE and BELONG ANYWHERE. The latter was the company's new slogan, introduced with much fanfare in 2014 along with its new curlicue logo, the Belo, which was widely interpreted as an abstract representation of part of the female anatomy.[2]

Such was Airbnb's grandiose sense of its own importance that in his keynote at the Airbnb Open, a gathering of hosts held for the first time in San Francisco in November 2014, Chesky recalled how his new head of global hospitality, Chip Conley—the only Airbnb exec better than the founders at slinging company agitprop—had predicted that the Airbnb community could win the Nobel Peace Prize within the decade. "I kind of laughed. I thought he was out of his mind," Chesky said in his speech. "And then suddenly you hear stories and you're like, 'We are not completely that crazy after all.' "[3]

Just like Uber, Airbnb was thoroughly infused with its founders' ambition and idealism, along with a generous seasoning of naïveté about how such musings might sound to the public. And like Uber, Airbnb in 2014 was a sponge for the optimism booming across Silicon Valley at the time. A few months before Uber raised its massive $1.2 billion Series D, Airbnb raised half a billion dollars from a group that included T. Rowe Price and two investors who had backed Uber, the private equity firm TPG, and Shervin Pishevar's new private VC fund, Sherpa Capital.

The astounding new valuation of the six-year-old company: $10 billion.

With 15 percent ownership stakes, Brian Chesky, Joe Gebbia, and Nathan Blecharczyk were each worth $1.5 billion on paper and joined the *Forbes* billionaires list the same year as Travis Kalanick, Garrett Camp, and Ryan Graves from Uber.[4] They were all in their thirties.

There was another, more unfortunate parallel between the companies. Like Uber, Airbnb was seemingly unprepared for the kinds of tragic events that were not only possible on its service, which discarded the safety protections found in conventional hotel rooms, but probably inevitable.

On December 30, 2013, only a day before young Sophia Liu was hit and killed by an Uber driver in San Francisco, Elizabeth Eun-chung Yuh, a thirty-five-year-old South Korean native from Ontario, Canada, died of carbon monoxide poisoning at an Airbnb in Taipei. She had traveled there with friends for a wedding and checked into an apartment downtown, where the landlord had recently enclosed an outdoor porch without properly venting the water heater or installing a carbon monoxide alarm.

According to a report in the *China Post,* her four friends in adjacent rooms were admitted to a local hospital and treated for carbon monoxide inhalation, but Yuh was found dead at the scene.[5] That night, her father, Deh-Chong Yuh, Tweeted at Brian Chesky:

Our daughter Elizabeth passed away in Taipei, Taiwan at a Airbnb arranged apartment from carbon monoxide inhalation on 30 Dec. 2013.[6]

Unlike the Sophia Liu tragedy, the incident received no attention in the Western media. When I later asked Airbnb about it, a spokesperson e-mailed me a statement: "We were extremely dismayed when we learned about this incident and we immediately reached out to the guest's family to provide our full support and express our deepest condolences. This was a tragic event and our focus has always been on supporting the family and taking action to help prevent this kind of incident from happening again. Additionally, we permanently removed the host from our community. Out of respect for our community members' privacy, we generally do not comment on the conversations we have with them."

The Yuh family, which did not respond to my attempts to reach them, contacted San Francisco personal injury attorney William B. Smith, who advised them to file a wrongful death suit and challenge Airbnb's fourteen-page terms-of-service agreement, which stated that hosts and guests assumed all risks and were responsible for adhering to local laws. But soon after, Smith told me, the Yuh family informed him that Airbnb had offered them two million dollars to resolve the matter. They decided to accept instead of filing a lawsuit.

According to a legal paper Smith later published on his law firm's website, Airbnb denied liability for the incident and specified that the settlement was "offered only for humanitarian reasons."[7] One attorney who worked for Airbnb later told me that Airbnb didn't have to settle the case but that, in such situations, Chesky focused intently on the right thing to do. To Smith, though, any hint of benevolence rang false. "People might pay money for humanitarian reasons, but corporations don't. They pay because of legal liabilities," he says.

Nearly two years later, journalist Zak Stone chronicled Yuh's death in a story about his own father's death at an Airbnb, which occurred when a tree branch attached to a tire swing fell on his head.

In a joint interview with all three Airbnb co-founders, I asked about such tragedies. "There is a certain statistical probability that extremely unlikely things will happen from time to time, given enough scale," Nate Blecharczyk answered. "It can be an opportunity actually to come out stronger. When something bad happens, we really look deep within and try to think hard about... what it is we can do going forward to make the service better."

Indeed, in the United States in 2014, Airbnb started giving away carbon monoxide detectors as well as first-aid kits, smoke detectors, and safety cards that advised hosts on emergency preparedness.[8] It also said that by the end of the year, hosts had to have smoke and carbon monoxide detectors in their homes, though there was no way to determine whether the hosts had actually installed them.

The Yuh tragedy epitomized the dueling realities facing Airbnb at the start of 2014. It wanted to be seen as an innovative hospitality brand bringing strangers together and providing authentic, intimate travel experiences. But it was also an internet marketplace that, like all such marketplaces, could not fully guarantee the scrupulous behavior of its hosts or the actual conditions its guests encountered.

The reality people saw often depended on where their sympathies lay. Regulators, left-wing politicians, hotel CEOs, union leaders, affordable housing advocates, and angry neighbors tired of carousing guests saw Airbnb as nothing but a rule breaker from the far-away land of arrogant, entitled billionaires. Investors, hosts, property owners struggling to make their monthly mortgage payments, travel-discount shoppers, and high-tech aficionados tended to believe in the startup with good intentions that was disrupting the stultified hospitality industry.

But despite its humble origins and more empathetic CEO, Airbnb was about to become every bit as controversial as Uber.

Steve Unger moved to Portland, Oregon, back in 2002 after losing his Silicon Valley job in the dot-com bust. With his husband, Dusty, he reinvented himself as the proprietor of the Lion and the Rose, a stately one-hundred-year-old Victorian inn with eight bedrooms, arched windows, a wraparound porch, and a turret on the third floor. In a good year, Unger catered to two thousand guests.

To register as a proprietor of a traditional bed-and-breakfast in Portland, Unger had to get a city permit that cost four thousand dollars, part of an ordinance that had been put into place to protect residential neighborhoods from too much commercial activity. So Unger was among those inclined to look askance when unlicensed Airbnbs started popping up across Portland, particularly since his business remained curiously slow even after the city emerged from the recession in 2012.

By the beginning of 2014, local hosts were imploring the city to reduce its B and B registration fees and stop the inconsistent, piecemeal enforcement of zoning laws that were shutting some Airbnb hosts down whenever neighbors complained to the city. Unger attended city council meetings on the issue and saw that Airbnb and its lobbyists were deeply involved in the debate. He marveled at the hosts that Airbnb reliably corralled to argue on its behalf. They offered sympathetic testimony about how renting out their spare rooms or in-law apartments on Airbnb allowed them to earn enough extra money to stay in their homes. Unger started calling this kind of arrangement "the good Airbnb." The bad Airbnb was made up of hosts with multiple properties and owners who didn't actually live in their homes for much of the year but rented them out online, keeping them off the housing market. These types of hosts weren't asked to testify at the meetings.

Despite resistance from neighborhood groups, Airbnb and its hosts succeeded in changing the law. Over the summer of 2014, Portland became the first city in the country to strike a deal with the company. The arrangement legalized short-term rentals in primary residences but limited unhosted rentals—that is, when the host was not present—to ninety days a year.[9] Registration fees were reduced from $4,000 to $180 and hosts were required to conduct a safety inspection of their homes, notify their neighbors, and register with the city. In return, Airbnb agreed to collect the 11.5 percent lodging tax on behalf of its hosts and send the revenue to the city (without including hosts' names and addresses).[10] The company also opened a customer-service call center in town.

There was peace in Portland, but Steve Unger didn't like it. "I believed that the ninety nights a year would be almost impossible to enforce unless Airbnb helped, and they never said they would," he told me. "They said it was critical to have ninety nights a year as one of the conditions of the agreement. They wanted people to go away on vacation and be able to rent their houses. And they make more money on entire place rentals."

For Airbnb, the Portland deal was one of the first steps in a new campaign to bolster its image and calm the mounting regulatory rancor. A blog post by Chesky published in tandem with the Portland announcement introduced a new initiative that the company called Shared Cities. It included Airbnb's pledge to make cities friendlier and nicer by, for example, helping hosts donate to local causes and matching their donations.[11]

At the heart of the proposal, Airbnb was offering cities a carrot, in contrast to the sharp stick Uber employed by weaponizing its customer base against its political opponents. Hotel taxes were the reward. Years earlier, Airbnb had said it shouldn't be responsible for collecting hotel taxes because it operated only as a marketplace.[12] Hosts, however, were unlikely to pay hotel taxes voluntarily. Airbnb now saw the advantage of conceding that point and facilitating tax collection itself in return for laws that sanctioned short-term rentals. "We're offering to cut red tape and to collect and remit taxes to the city of Portland on behalf of our hosts," Chesky wrote. "This is new for us, and if it works well for our community and cities, we may replicate this project in other U.S. cities."[13]

That, it turned out, was foreshadowing. A week later, the company said it planned to start collecting the 14 percent hotel tax (aka the transient occupancy tax) in San Francisco[14] and even agreed to pony up tens of millions in unpaid back taxes (it never specified the exact amount).[15] Over the next year, it would strike taxes-for-legalization agreements in Chicago,[16] Washington, DC,[17] Phoenix,[18] Philadelphia,[19] and elsewhere. Amsterdam became the first city in Europe to sanction short-term rentals, permitting residents to rent their homes for no more than two months a year and only to four people at a time.[20] France also legalized short-term rentals of primary homes and empowered its cities to pass additional restrictions on rentals of nonprimary residences.[21]

Chesky was optimistic that Airbnb had turned the tide when we discussed the issue in 2015. "Every city used to look to New York to figure out what to do," he told me. "Now I think cities are deciding, we are going to figure out what's best for us."

But New York City was still a flash point, and it was here, in one of the company's largest markets, that Airbnb first underestimated the powerful political forces that were beginning to mobilize in the wake of its success. In the spring of 2014, Airbnb was negotiating with the office of New York attorney general Eric Schneiderman to end the long standoff in New York. The parties came close to reaching a settlement that would resolve the AG's issues with the company and authorize it to collect taxes on behalf of its hosts, according to three people familiar with the discussions. But then, in an abrupt turnaround, the city refused to finalize the agreement. Almost overnight, Airbnb had somehow become politically radioactive in New York City.

People involved in the discussions say there were two reasons for the reversal. Airbnb had just completed the round of funding that valued it at $10 billion. It was now worth more than major international hotel chains like Hyatt Hotels and Wyndham Worldwide. Jolted by this, these companies suddenly woke up to the looming threat in their midst. Ten days after the news, the American Hotel and Lodging Association, the primary trade group for the 1.9 million employees in the U.S. hospitality industry, released a statement announcing that it would begin tracking Airbnb and other short-term rental sites and drawing attention to issues like taxes, adherence to disability laws, the protection of residential areas, and the preservation of neighborhood parking.[22]

At the same time, Airbnb had reached out to the New York City chapter of the Service Employees International Union in hopes of giving its hosts access to certified unionized housecleaners who could be summoned on demand. That alienated another hotel union, the powerful Hotel Trades Council, which feared such an agreement could further legitimize Airbnb. It too spun up a campaign to curb short-term rentals, funding the creation of a New York–based lobbying group called Share Better.[23]

Airbnb now had two significant enemies: hotels and their powerful employee unions. Both were well organized and deep pocketed, and both had strong relationships with local governments. A lawyer working for Airbnb on the attempted settlement in New York City said that within twenty-four hours, the hotel unions and their representatives scuttled any prospective deal, insisting that the city should not do anything to legitimize Airbnb. "At that moment everyone panicked and said, we are not touching this for a while," this person said.

Airbnb attempted to swing popular opinion to its side but didn't have the same populist tools as Uber. Hosts who rented out their apartments only a few times a year were unlikely to show up for protests at city hall at three o'clock in the afternoon. So that summer, Airbnb hired Bill Hyers, the manager of de Blasio's successful campaign for mayor, and he placed ubiquitous ads in New York City subways featuring photos of smiling New Yorkers using the service to make ends meet. ("The good Airbnb," according to Steve Unger.) NEW YORKERS AGREE: AIRBNB IS GREAT FOR NEW YORK CITY, read the tagline on the subway posters. Many of these posters became the target of graffitied scrawls such as *Airbnb accepts NO liability* and *The shared economy is a lie.*[24]

By the end of 2014, the chances for a political agreement in New York looked dim. As the report that fall by New York attorney general Eric Schneiderman revealed, more than two-thirds of

Airbnb rentals in the city violated the draconian Multiple Dwelling Law—the law that Toshi Chan had protested back in 2010 that said New Yorkers couldn't rent out their homes for less than thirty days. And 6 percent of hosts in the city were renting out multiple units on Airbnb, generating 37 percent of the company's revenue there.[25]

Momentum was shifting elsewhere too. The laws passed in the Shared City agreements with cities like Portland were being flouted, just as Steve Unger had feared. They required hosts to register with their cities, but, despite the fanfare that accompanied the deals, few did. In the face of this, Airbnb refused to put in place restrictions to force compliance—for example, requiring hosts enter valid registration numbers or preventing a host from listing multiple properties. In interviews, company executives pointed out that law enforcement was not typically the domain of a private company and complained that the registration process was often too complicated and time-consuming (hosts in San Francisco had to make an appointment and then appear at city hall in person to present valid documents[26]). But it was also impractical for the cities to police the thousands of anonymous people using the home-sharing site, and Airbnb didn't seem all that eager to help.

Airbnb had said it wanted to talk candidly with cities, to play by the rules, to be a partner. But in the end, there emerged an unavoidable fact: Chesky was every bit the warrior Travis Kalanick was. He believed so much in the promise of his company that he was going to fight for every inch of territory.

In July 2015, I traveled with Chesky and half a dozen of his colleagues to Nairobi, Kenya, for a conference called the Global Entrepreneurship Summit, held annually since 2010 by the U.S. State Department. It was a celebration of innovation and private enterprise, and Chesky, as a Presidential Ambassador for Global Entrepreneurship (PAGE), would be meeting with President Obama, speaking on a panel, and talking to African entrepreneurs. Bonus for Airbnb: it was neutral territory thousands of miles away from Airbnb's nearest regulatory skirmishes in Europe and the United States.

The trip was long and security in Nairobi was tight. The Airbnb entourage stayed not with one of the city's 788 Airbnb hosts, as they usually would, but at the Fairmont Hotel, its front driveway ringed by metal gates and security checkpoints. The capital was on lockdown for Obama's first visit to Kenya since he had become president. Soldiers holding automatic weapons lined the streets from the airport to the city while residents crowded at intersections, straining for a glimpse of Obama or other visiting dignitaries. Billboards everywhere bore his image along with the words WELCOME HOME, PRESIDENT OBAMA!

It was a historic moment but also an opportunity to see Chesky the diplomat, away from the regulatory threats and operational challenges that confronted him at home. His performance was impressive. Few Silicon Valley execs can so effectively phase shift—digging into operational complexities at one moment, negotiating with politicians the next, and then leaving it all behind to speak in relatable tones to students, other startup founders, and the general public. Chesky did this with ease, and it was a reminder of the remarkable personal skills that had propelled his company to such astounding heights.

Chesky and other PAGE ambassadors met with Obama privately on the morning of the conference. Obama, I later heard, gave him a "bro hug" and made a reference to news of a new Airbnb fund-raising round (new valuation: $24 billion). "It looks like you are doing very well," he said, revealing that even POTUS was watching the upstarts. Later, after Obama's moving speech about his family's history in Kenya and the opportunities to spur economic development in Africa, the president advised a local Kenyan founder who had joined him onstage, "We have the founder of Airbnb here. You can talk to him, he's doing pretty good." The packed audience laughed.

On a panel later in the day with five other tech-company CEOs, Chesky was easily the most charismatic speaker. He worked from a reliable playbook, mining the early history of Airbnb to offer business lessons and inspiration: "Not too many years ago I was an unemployed aspiring entrepreneur, living in an apartment with my roommate Joe. We couldn't figure out how to make rent, and one day a design conference came to San Francisco. All the hotels were sold out. We thought, *What if we could turn this house into a bed-and-breakfast for the conference?* We called it the AirBed and Breakfast."

All startup stories bend toward fable. Airbnb's had evolved into an oral history to be recited in keynotes, at new employee orientations, and probably around the campfires at company retreats. "When I started Airbnb with Joe and Nate, I looked at successful entrepreneurs and I did not see myself in them," Chesky said. "I thought they were put on a pedestal. They seemed smarter than me. More successful."

The next day we drove to a startup incubator called iHub, twenty miles west of downtown. Even in subtropical Africa, startup founders sought places to cluster, work, and strategize about how to surf the great wave. More than two hundred people had jammed themselves into a common room on the fourth floor, where the air started out humid, moved quickly to sweltering, and then progressed to suffocating. Chesky, wearing a tight gray T-shirt with the Belo logo on the front, didn't seem affected and spoke for ninety minutes straight. "During the financial crisis, people started using Airbnb to stay in homes. For us, it was a turning point. Fast-forward to today, six years later, we have one point five million homes around the world. That's as many rooms as Hilton and Marriott combined. This summer on a peak night, we'll have close to one million people in a single night staying in a home." The crowd, delighted to be in the presence of Silicon Valley royalty, erupted in rapturous applause.

There were many questions. A Kenyan in a yellow jacket stood up and asked about Airbnb's regulatory issues. Apparently Nairobi wasn't that far from the company's legal scrums after all.

Chesky's answer was revealing and optimistic—perhaps overly so. "When there's a cool new business on the internet, that's great," he said. "But when the internet moves into your neighborhood, into your apartment building, and you don't know anything about it, suddenly people assume the worst and they have a lot of fears.

"So there's a couple of things you need to do. The first thing you need to do is grow really, really fast. You either want to be below the radar or big enough that you are an institution. The worst is being somewhere in between. All your opposition knows about you but you are not a big enough community that people will listen to you yet.

"You have to get to what I guess I'd call escape velocity. If a rocket takes off, there's a bumpy ride before you get to orbit, and then there's a little bit more stillness.

"The second thing is you need to be willing to partner with cities and tell your story. We found the most important thing to do is to go and meet city officials. If people dislike you or if people hate you, it's often normal to ignore them, to avoid them or to hate them back. The only real solution is to meet the people that hate you. There's an old saying that it's hard to hate up close. I have found that. It's really hard to hate somebody when they are standing right in front of you."

Two years before the proliferating fights over hotel taxes and host registrations, Peter Kwan started renting out the sunlit spare bedroom on the ground floor of his charming Edwardian home in San Francisco's North Beach neighborhood. His longtime roommate had just moved to Germany, and Kwan, in midfifties, lived alone with Haley, his excitable West Highland white terrier. He was semiretired from teaching constitutional law and eager to meet new people and keep his house—the spare room was great to have available for visits by his sister and nephew. So he decided to give Airbnb a try.

Airbnb exceeded Kwan's expectations on all counts. Over the years he met travelers from dozens of states and countries and stayed in contact with many of them. Using Airbnb was "better than I ever imagined or hoped for," he says. "It's been both satisfying emotionally and rewarding economically."

But since Kwan was trained as a lawyer, after a few months of hosting, he started to wonder: Did he have liability insurance if a guest got injured? Should he collect the city's transient occupancy tax? Was all this even legal? He checked the Airbnb website and there were no answers. The startup, he reasoned, would need a sizable law firm to educate hosts about the varying ordinances in all of its thousands of cities and countries. So Kwan did some research himself. Back then the answer to the last question, it seemed, was technically no, at least in San Francisco, since bed-and-breakfast purveyors needed to register and pay various fees, just like they did in Portland. But, of course, the law wasn't being rigorously enforced.

Kwan decided to gather a group of hosts together to share information and navigate the emerging complexities of the so-called home-sharing economy. He announced the formation of his club on Craigslist and held the very first meeting of the Home Sharers of San Francisco in his living room in 2013. The group would eventually attract twenty-five hundred members. Seeking to avoid any conflicts of interest, Kwan decided the group would not allow Airbnb employees or city or state government workers to join.

Kwan's group got so large that eventually it had to start gathering in public libraries instead of living rooms. They shared hosting tips, talked about issues like insurance, and swapped stories of nightmare guests (always the most enjoyable discussion). Then things got serious. In the wake of Airbnb's agreement to collect hotel taxes, the city's board of supervisors was considering legalizing short-term rentals. The Home Sharers lobbied to keep the names and addresses of hosts private and to maximize the number of nights they could rent out their properties each year.

The bill was written by board president David Chiu, a longtime political ally of Airbnb investors Ron Conway and Reid Hoffman.[27] The bill passed in October 2014 and became law the following February. Under the new law, hosts were allowed to rent their homes for under thirty days without restriction when they were present and for ninety nights a year when they were absent. Hosts were also required to register with the city and to carry liability insurance, and the city agreed to create a new Office of Short-Term Rentals to administer and enforce the law.[28] Mayor Ed Lee signed the bill, and Airbnb celebrated it in a blog post, commending it for containing "sensible rules of the road" and as "a great victory for everyone who wants to share their home and the city they love."[29]

While it might have seemed like a victory, this was actually the beginning of Airbnb's next fight.

At the time, San Franciscans seemed increasingly conflicted about the technology renaissance in their midst. The city that celebrated its bohemian past and distinctive neighborhoods was at the nexus of several converging trends: the acceleration of the internet economy, the migration of Silicon Valley startups up Highway 101 and into the city, and the infusion of millennials into cities. Home prices in the city were skyrocketing as a result, and gentrification was rapidly changing beloved neighborhoods, such as the predominantly Latino Mission District.

It all produced a kind of poorly articulated rage. The convenient culprits included the street-clogging double-decker company buses that ferried employees to the offices of Google, Facebook, and Apple; the tech companies themselves; and the so-called tech bros, vaguely defined stereotypical males who could be relied on to regularly Tweet or blog something racist, sexist, or generally insensitive, thereby indicting the entire tech industry. "Number 5: the 49ers," wrote a startup founder named Peter Shih in a much reviled blog post titled "Ten Things I Hate About You, San Francisco Edition." "No, not the football team, they're great. I'm referring to all the girls who are obviously 4's and behave like they are 9's."[30]

Another handy scapegoat was Airbnb, which was having an undefined but real effect on the number of rooms and homes for rent in San Francisco and other cities as landlords like Peter Kwan opted to market their spare bedrooms to tourists instead of renting them or selling their homes to permanent residents.

The issues facing San Francisco pitted new residents against old, techies against nontechies, and centrist Democrats against progressives. Airbnb was a tempting wedge issue in this fight, a way to muster opposition to the tide of gentrification under the banner of a universally appealing word: *affordability*. Even though the new Airbnb law was only a few months old, opponents tried to get the legislature to strengthen its restrictions. When that failed, they got fifteen thousand signatures to put a new initiative, called Proposition F, on the ballot in the fall 2015 elections, when progressives were looking to recapture the board of supervisors and unseat Ed Lee, the moderate Democratic mayor.

Prop F attempted to reduce the number of rentable host-absent days a year from ninety to seventy-five, to make renting out entire in-law apartments illegal, and to allow citizens to sue neighbors who lived within one hundred feet and were violating the law.[31] These were tough measures that threatened to swamp the city in a caustic volley of lawsuits between neighbors. The initiative's proponents, a trio of local activists backed by the city's tenants' union and apartment association, argued that the original law had no teeth, in part because hosts were unlikely to voluntarily register and the city did not have the resources to make them do so.

Peter Kwan and his fellow hosts mobilized against Prop F. They started a separate group, called the Home Sharers Democratic Club, to give hosts a political voice in the fray, convened press conferences, and organized phone campaigns to educate the citizenry on the folly of the initiative. "We were made a scapegoat for the housing crisis," Kwan told me over home-cooked Singapore-style noodles in his dining room that year, which I ate while Haley nipped at my ankles. "Yes, we do have a severe housing shortage and affordability problem, and yes, home sharing probably does play some part in contributing to the severity, but I don't think anybody really knows the degree of the contribution."

Airbnb also mobilized against the initiative. San Francisco contributed only a small percentage to its growing global business but as the place of its birth and its hometown, Airbnb felt the symbolic stakes were high. It backed an organization called San Francisco for Everyone and contributed more than eight million dollars to the campaign. That fall the group plastered the city with NO ON F posters, outdoor billboards (WHICH NEIGHBOR IS GOING TO SNITCH ON YOU?), and ubiquitous radio and TV ads featuring members of "the good Airbnb," like an elderly couple who lived on a fixed income and testified sweetly that "home sharing is helping us stay here."

Meanwhile, the pro–Prop F forces trotted out tenants who had been booted from their homes by avaricious landlords eager to use short-term rentals to boost their earnings ("the bad Airbnb")[32] and plastered posters around town that said FIX THE AIRBNB MESS. A few days before the vote, seventy-five protesters banging drums, blaring horns, and chanting "No more displacement in this city!" occupied the atrium in Airbnb's swank headquarters. They spent ninety minutes there, delivering angry speeches and releasing clusters of black helium balloons that carried posters with words like *evictions* and *deregulation* to the atrium ceiling. Airbnb employees watching the commotion from the third-floor balcony got a good look.[33]

The polling data showed a tight race, with the pro-Airbnb forces commanding a narrow lead. Chesky stayed away from the scrum but later discussed the stakes for the company. "You can win ten cities in Europe but you lose your home city and it basically just seems like you're losing ground completely," he said on a technology podcast. "This was a big, big-time fight."[34]

Peter Kwan felt the outcome was in doubt until Election Day, November 3, when voters rejected the proposition by a surprisingly wide margin, 55 to 45 percent.[35] Airbnb had won. Kwan and other members of the Home Sharers gathered at the Oasis Nightclub to revel in their victory, but for several reasons, Airbnb itself wasn't exactly exulting in the moment.

A few weeks before Election Day, Airbnb advertisements appeared on billboards and at bus stops around the city bragging in a cheeky tone about the impact of its tax-collection efforts. DEAR PUBLIC LIBRARY SYSTEM: WE HOPE YOU USE SOME OF THE $12 MILLION IN HOTEL TAXES TO KEEP THE LIBRARY OPEN LATER, read one. Suggested another: DEAR BOARD OF EDUCATION: PLEASE USE SOME OF THE $12 MILLION IN HOTEL TAXES TO KEEP MUSIC IN SCHOOLS.

The advertising agency TBWA\Chiat\Day had been hired by Airbnb to promote its tax-collection efforts, but the campaign they came up with was widely ridiculed on Facebook, on Twitter, and in the national media for being patronizing, ill timed, and just plain baffling. Airbnb appeared to be arrogantly patting itself on the back for something it should have been doing regardless. After the backlash, Airbnb quickly pulled the campaign and apologized for it. Chesky later said that he had not seen or approved the ads. But the damage was done. With Proposition F, Airbnb had been fitted for the black hat. Though it won the battle, in an act of inexplicable self-sabotage, it had slipped the hat on itself.

After the vote, Chesky called an all-hands meeting and invited select hosts from around the city to attend. Peter Kwan was there, with Haley in tow. Employees and guests gathered in the fifth-floor cafeteria. Chesky and Gebbia spoke, as did Jonathan Mildenhall, the chief marketing officer, who took responsibility for the ads and apologized to the company. Kwan recalls that some employees were close to tears, upset not just by the ads but by all the rancor over Prop F and the way the company had been portrayed in the media. "I think there was a certain sense of betrayal," he told me. "The whole Prop F debate made a lot of people feel uncomfortable about what they were doing. It came at this time of constant barrage of criticism of the company that just opened up the floodgates of emotion." Chesky, looking distressed, "didn't pull punches. He said, 'We screwed up,'" Kwan recalls.

Airbnb staggered away from its costly Prop F victory in other ways as well. Mayor Ed Lee had won reelection and Proposition F had been defeated, but progressive Democrats took control of the board of supervisors and would pass even more draconian anti-Airbnb legislation in 2016.

San Francisco had been a harbinger, a taste of things to come in other cities in the country and around the world. The unlikely political coalition forged to counter Airbnb in its hometown was taking shape in Portland, Los Angeles, Chicago, Boston—and on and on. Anticipating this, Chris Lehane, once a hardball political operative for President Bill Clinton and Vice President Al Gore and now Airbnb's new head of global policy and communications, held a press conference after the election. He announced Airbnb would finance the creation of a hundred grassroots political clubs to advocate for home-sharing. "We're going to use the momentum of what took place here to do what we did in San Francisco around the world," Lehane said.

Airbnb had not, as Chesky had hoped in Kenya, achieved any kind of "escape velocity." In fact, it was looking like the company had unleashed political forces that were not going away anytime soon.

Less than a week after the Prop F vote, Chesky and six hundred of his employees traveled to Paris for the second Airbnb Open, held in the cast-iron-and-glass Grande Halle at the Parc de la Villette, a picturesque park lined with canals in the nineteenth arrondissement. Here was another abrupt shift between the company's dueling realities. Within the span of just a few days, it had moved from navigating the muddy trenches of local politics to the happy self-absorption of its annual community festival.

Five thousand hosts from a hundred and twenty countries paid three hundred dollars a ticket for the three-day event, which was thoroughly infused with the spirit of a tent-revival meeting. Speakers embraced each other onstage and led the crowd in chants and dances. A Cirque du Soleil performer erected an impressive edifice of balancing sticks in her outstretched hands. There were luminaries like Alain de Botton, the Swiss author and philosopher, who said, "Psychological hospitality trumps material hospitality every single time." The crowd stood and cheered repeatedly during the event, responding to rousing proclamations ("You are truly revolutionaries!"), as if the speakers were blowing dog whistles.

Occasionally the audience was yanked back to the other reality. "This generous idea is growing in Paris," said Jean-François Martins, deputy mayor in charge of tourism, on the first morning. "But big ideas need some regulation to protect them from people who want to use it in a not very generous way." Chris Lehane also appeared onstage and spoke to the gathered hosts as if they were infantry in the French marines. "We are going to have more fights and we are going to have more battles in the days, months, and years to come," he said. "When this community is empowered to be a movement, we cannot be beat."

In the middle of it all were the three founders, Chesky, Gebbia, and Blecharczyk, now the billionaire gurus of a peculiar internet sect. They spoke together and, in addition, Chesky and Gebbia spoke individually; they answered questions from the crowd and revisited Airbnb's fabled origins. Gebbia's speech was the most memorable. He appeared onstage wearing a wool hat, gloves, and a scarf, with two colleagues tossing fake snow on his head. It was an eccentric piece of performance art meant to re-create his early trips to solicit hosts in New York City during the winter of 2009.

During his speech, Chesky announced a new initiative, the Community Compact, the successor to the Shared Cities program from a year before. The compact committed the company to booting illegal hoteliers off the site, paying hotel taxes, and publishing anonymized data in its largest markets, including information on the percentage of Airbnb hosts who were sharing their permanent homes. "It's not a new commitment but people didn't believe us so we decided to just say it again and write it down," Chesky later said.

The abundance of company worship at the Airbnb Open was tough for any hardened journalist to handle. But the hosts themselves, walking around the Grande Halle and attending speeches and seminars with whimsical names like "Hospitality Moments of Truth," were disarming and inspiring. They were Airbnb's most persuasive evangelists.

Here was a group that loved the company and what it stood for, demonstrating a kind of loyalty and passion that Uber, for example, would never see from its drivers. Among the hosts I met that week were Tanny Por, a so-called superhost who rents out a spare bedroom in her home in Nuuk, the capital of Greenland. Por moved from Australia to Greenland in 2013 when her husband got a new job. Renting a room in their home was a way to meet people but also a kind of social lifeline, allowing them to stay connected to the cosmopolitan world from the icy shores of the North Atlantic. "We spend a whole lot more time with our guests than everyone else, simply for the fact that there's not so much to see or do in Nuuk," she told me.

I also met Julia de la Rosa and Silvio Ortega, part of a small contingent of hosts from Cuba, which, to great fanfare and copious amounts of media attention, Airbnb had recently opened to travelers from the United States. Since they'd lost their jobs in the country's economic collapse in the early 1990s, Ortega and de la Rosa had run a bed-and-breakfast in Silvio's ten-bedroom family home in a Havana suburb. Until Airbnb, they could solicit guests only via travel agents and random online bulletin boards. On Airbnb they could see their guests' profiles and post photos and facts about their home so that visitors weren't surprised when they arrived.

Since they started listing on Airbnb, the couple had hosted several dozen groups of Americans, including some university students and their professor. The biggest challenge, they found, was that American tourists all wanted to talk to them at length. "They are so friendly and open and just want to understand Cuba. It's amazing," Julia said.

There was something authentic and charming about the Airbnb community, I found that week. Maybe I was drinking the Kool-Aid, and certainly Airbnb served up generous amounts of it, but in that context it was difficult not to feel sympathetic toward a service that was allowing people to experience the world through one another's eyes.

That overarching impression was especially hard to shake after the events of the second night of the conference, November 13, 2015, when terrorists, in a coordinated attack, killed 130 people at a soccer stadium, several cafés, and the Bataclan music theater in Paris and its suburbs. I was dining less than a mile away from the Bataclan when the attacks occurred. Chesky, Gebbia, and Blecharczyk were eating dinner with their families and forty other long-term company employees at a local Airbnb.

Everyone had to stay in place for hours as sirens and frantic activity filled the fall night. Chesky later recalled heading to the master bathroom, seeking privacy so he could coordinate a response with his security team and with Belinda Johnson, who was with another group of employees at a nearby restaurant. Together they made sure the company checked on every employee and host in Paris. Everyone was accounted for. Later that night, the company canceled the third day of the event.

I Ubered home that night to my Airbnb, near the Cathédrale Notre Dame de Paris. Once back, I received a frantic phone call from my worried host, Ivan, whom I had never met in person (he was out of town and had left me the keys) but who wanted to make sure I was okay. The next morning, he e-mailed: "I was relieved to hear you yesterday on the phone. I hope you are well today despite the critical times in Paris." He invited me to stay for as long as necessary until the travel situation in the city returned to normal.

It was no doubt one of many simple acts of kindness that week and perhaps the sort of unquantifiable variable the founders believe should be considered in the grand political calculation of Airbnb's impact in a dangerous world.

There were more battles, small victories, and meaningful setbacks for Airbnb, all adding up to reliable drama. In 2016, the city of Berlin made it illegal to rent entire homes and apartments for short periods and asked citizens to anonymously report violators. Offenses were punishable by fines of up to one hundred thousand euros.[36] That same year, Tokyo battled over Airbnb and considered draconian restrictions for the new phenomenon of house rentals, called *minpaku* in Japanese. In a moment of surprising candor, one lawmaker revealed to Bloomberg's Yuji Nakamura why the city was considering restricting the twenty-six thousand Airbnb hosts in the country. "The hotel industry had very serious concerns, so we set the minimum number of nights at a level that lowers the chances for competition," he said.[37]

In June 2016, San Francisco's progressive city council passed another ordinance that would fine Airbnb anytime a host violated the local law. Airbnb promptly sued in federal court, arguing that the bill violated an internet statute that protects websites from being held liable for any content posted by their users. When it appeared it would lose the case, Airbnb caved and agreed to make sure hosts register with the city and limit unhosted rentals to 90 nights a year. Also that month, the New York State legislature passed a bill that would slap fines as high as $7,500 on anyone in New York City caught listing an entire property on Airbnb without a host present for less than thirty days. The company argued that it was an indiscriminate bill that lumped the good Airbnb with people trying to exploit the service. On October 21, New York governor Andrew

Cuomo signed it into law, and Airbnb filed another lawsuit in federal court against attorney general Eric Schneiderman, Mayor Bill de Blasio, and the City of New York.

If the political tide was turning against Airbnb, Chesky hadn't seemed worried. "We're in thirty-four thousand cities, so this experiment is being played out all over the world," he told me that July. "We have tax agreements with over a hundred and sixty cities around the world, and I think it's pretty clear that this is an idea that is here to stay."

As it had promised, Airbnb released reports about the statistical makeup of its community in major cities. Several times, it ousted large groups of hosts who were renting multiple properties from the site. Some saw this as a good-faith effort to tailor its business to the housing realities of big cities. But opponents accused the company of booting illegal hoteliers to create a more favorable picture of its data and they questioned Airbnb's commitment to dealing candidly with regulators.[38] "I haven't seen them demonstrate anything other than trying to maximize their revenue," said Murray Cox, a "data activist" and creator of the website Inside Airbnb, which scraped the Airbnb website to collect independent data on hosts.

In May 2016, Gregory Slenden, an African American from Richmond, Virginia, filed a civil rights lawsuit in Washington, DC, against Airbnb for ignoring his complaint that he had been subjected to racial discrimination on its site.[39] There was academic backing for his claims; Ben Edelman, an associate professor at the Harvard Business School, had published two studies showing that Airbnb users were statistically less likely to host or stay with minorities.[40]

Slenden's charges sparked an uproar. Using the hash tag #Airbnbwhileblack, African Americans took to social media to share their own experiences of encountering prejudice on the home-sharing site. Many testified that rooms would suddenly become unavailable or hosts would simply fail to reply when they tried to book accommodations. The U.S. media dug into the issue (the *New York Times:*

"Does Airbnb Enable Racism?"), and the verdict, for Airbnb, was unfavorable.[41]

The fracas challenged some of the company's most sacred ideals. It was supposed to eradicate the biases of the past, not give them new life; user photographs were supposed to help establish trust, not provide an opportunity for racist judgments. For once, Chesky didn't have an easy answer, since Airbnb had little control over the personal choices of its hosts or guests. Dismayed, he hired former U.S. attorney general Eric Holder and Laura W. Murphy, a former director of the American Civil Liberties Union's Washington, DC, office, to develop a way to combat discrimination on the service.[42] In September 2016, the company released a thirty-two-page plan for addressing the issue; among other things, Airbnb pledged to minimize the prominence of user photographs and asked hosts and guests to agree to a policy of nondiscrimination. "I think we were late to this issue," he said that summer. "Joe, Nate and I, three white guys—there's a lot of things we didn't think about when we designed this platform."[43]

Inside Airbnb, life was defined not only by these engrossing external conflicts and controversies but also by the frantic rhythm of relentless growth. On New Year's Eve 2015, it had booked 550,000 guests; on New Year's Eve 2016, it was 1 million; and by the middle of 2016, it was booking 1.3 million a night.[44] All of its internal graphs bent upward and to the right. The company was warping the gravitational field of the hospitality industry. To keep up with Airbnb, the online travel giant Expedia bought HomeAway, a rival vacation-rental service, for $3.9 billion.[45] And in 2016, hotel rates in New York City were the lowest since the great recession. Some industry observers blamed the new competition.[46]

There were new investors, bigger valuations, more employees. By mid-2016, Airbnb had twenty-six hundred workers. More than half had joined in the previous twelve months. Departments doubled or tripled in size, disrupting any hope employees had of finding a regular rhythm. One Airbnb employee told me her team had been shuffled four separate times and given four different bosses in the course of two years. Meanwhile, Airbnb's new chief financial officer, Laurence Tosi, the former CFO of the Blackstone Group, imposed new controls on what had been an undisciplined spending culture. For the first time, departments were required to come up with rigorous annual budgets and head-count projections — and to stick to them. Several employees described the company to me in late 2016 as being less fun and entrepreneurial.

These were the unmistakable signs of a startup growing up, shedding its identity as an upstart, and marching toward an eventual IPO. Just like Uber, it would first have to convince public investors that it had resolved its regulatory problems and achieved Chesky's coveted escape velocity. Beyond that lay the ceaseless drumbeat of corporate adulthood.

CHAPTER 12

GLOBAL MEGA-UNICORN DEATH MATCH!

Uber versus the World

Transportation as reliable as running water, everywhere and for everyone.

—Uber mission statement

Uber too sat on the threshold of corporate maturity. But first, it had to get through its final years of awkward adolescence—stormier, more adversarial, and even more eventful than the ones navigated by Airbnb. Uber was a gangly athlete who grew a foot in the span of two years and no longer fit into his clothes—and who had problems with aggression. It was, quite frankly, a marvel to behold.

By the beginning of 2014, Uber had introduced the ridesharing service UberX in twenty-eight cities. By the end of 2016, UberX and other flavors of Uber ridesharing were in more than four hundred and fifty major cities around the world. Allowing nonprofessional drivers to pick up passengers in their own vehicles had become a global phenomenon, bringing a new type of flexible work to drivers, lowering the price of transportation, and changing the way people traveled around cities.

Propelled by lower prices and higher ride volumes, Uber's already vigorous business exploded—exponentially. Uber booked a cumulative two hundred million rides by the beginning of 2014, one billion rides by the beginning of 2016, and two billion rides only six months later. Its employee base grew from 550 to 8,000 during this time. A $1.4 billion investment led by Fidelity Investments and BlackRock in June 2014 valued the company at $18 billion. That had already seemed bonkers then, but two years later, its valuation had more than tripled, to $68 billion, making Uber the most highly valued privately held technology startup in history.

The rise of ridesharing also sparked another wave of conflict in nearly every major city and country in the world. Travis Kalanick had promised a more optimistic and mature style of leadership during Uber's PR calamities of 2014, but while he toned down his rhetoric, he didn't modulate his ambitions. And that engendered more regulatory combat and fierce competition than any technology startup had ever seen.

London was one of the first European cities to grapple with the disruptive implications of ridesharing. The city had a proud taxi heritage, with drivers of its iconic black cabs required to study the city's demanding street grid for three years to pass a test grandiosely referred to as "the Knowledge." London's cabbies were proud and skilled, and they earned a comfortable middle-class wage. They were also furiously resistant to any kind of change other than fare increases, which they championed with periodic appeals to the regulator that set their fares, Transport for London (TfL), and backed up with threats of taxi strikes.

When Uber first came to town, in 2012, the black-cab drivers were already feeling threatened by the swell of minicabs and four-door sedans that could be booked by telephone or in person at a minicab office. Minicabs were legally prohibited from running taximeters or picking up passengers on the street.

Uber blew away all the distinctions between the black cabs and minicabs, while the ubiquity of GPS rendered the Knowledge essentially superfluous in an unkind instant. Then it introduced a version of UberX in London in June 2013, after operating its luxury car service there for a year. Like minicab operators, UberX drivers would have to obtain a private-hire license and commercial insurance. They did not have to pass the Knowledge. Unlike minicab drivers, UberX drivers would respond to requests from the Uber app and wait for rides while they were on the road. Uber did not ask for permission before proceeding with UberX; it simply judged that fifty-year-old regulations had not accounted for new technology such as electronic hails.

UberX operated quietly in London, at first. But the black-cab drivers were attuned to new competition and did not tolerate it graciously. On June 11, 2014, they objected to the city's embrace of UberX by gridlocking the city with a midday strike, blocking traffic on Lambeth Bridge over the Thames and paralyzing the city center.[1]

"All the years I've been driving a cab I had people fighting over me. Now it's a miracle if I get a job," John Connor, a black-cab driver from the East End with forty-four years of experience, said as he ferried me from Heathrow Airport into Shoreditch a few months later. He was one of the ten thousand taxicab drivers who had participated in the strike. "We had to let people know that they can't shit on us!" he said.

Many of the new UberX drivers, he noted, were immigrants from countries like Pakistan, Bangladesh, Somalia, Ethiopia, and Eritrea. They were happy to work eighteen-hour days for below-minimum wage. He had a family to support! These were the same issues—immigration, globalization, and middle-class anxiety—plaguing all Western countries in the early twenty-first century. "I've never seen a change like this in my life. The game is finished," he said as the car idled in the city's stalled traffic.

Uber said sign-ups jumped 850 percent after the strike. Drivers had inadvertently brought Uber more attention. Seven thousand people were driving for the company in London by that fall. At the same time, Jo Bertram, Uber's steely London-based regional manager, was subject to such vicious online hostility that she was forced to abandon social media. She had repeatedly faced the combative British press, but the vitriol on Twitter was too much. "There was a barrage of abuse," she told me. "My friends said, 'Stop reading this. It's not healthy.' We just handed it off to a colleague."

The antagonism toward Uber in London simmered that year as UberX grew more popular. In 2015, swamped with mounting driver complaints, the embattled TfL moved to suppress the Uber insurgency. It proposed rules that, among other restrictions, would prohibit Uber from showing available cars in its app and require drivers to wait a minimum of five minutes before picking up passengers who had solicited a ride.[2] These were irrational measures, mostly meant to curtail Uber's appeal. Emotions were running high. "That Travis, he's so smarmy, I could just punch him," Steve McNamara, general secretary of the Licensed Taxi Driver Association, a black-cab trade group, told me on my visit.

At the eye of the storm was London's mop-topped Conservative mayor, Boris Johnson, who would later rise to international attention as a main backer of Brexit, Great Britain's withdrawal from the European Union. Johnson was in a tough spot. He had solicited the support of black-cab drivers during his mayoral run in 2008, even printing his campaign slogan on taxi receipts.

At first Johnson noted that Uber was systematically breaking minicab regulations by allowing its drivers to cruise the streets and wait for passengers. But he also observed that technology had wiped away any rationale for the distinction between black cabs and minicabs. At an open question-and-answer forum in September 2015, he called the black-cab drivers who had packed the auditorium "luddites who don't want to see new technology." The drivers rose in a chorus of angry jeers, creating pandemonium in city hall and getting them ejected from the building.[3]

But just like in U.S. cities, Uber held the dominant hand in London because residents loved the service. The company mobilized not only an army of practiced lobbyists but two hundred thousand customers who put their signatures on a petition calling for the TfL to drop its proposed restrictions. In January 2016, it did. Johnson conceded that the regulations "did not find widespread support" and said that lawmakers could not "disinvent the internet."[4]

Boris Johnson's counterparts in continental Europe did not necessarily agree with him. In France, the institutional reflex against Uber was strong. In early 2014, with Uber growing in Paris—the city had become the startup's sixth market more than two years before—the French legislature ruled that drivers had to wait fifteen minutes before picking up a passenger that hailed a car on the Uber app. A French administrative court overturned the ruling but it was indicative of the fight to come and of the influence wielded by the country's two largest taxi companies, which had consolidated control of the taxi market.[5]

At that point Uber was using only professional drivers in France. But a private chauffeur's license cost three thousand euros and there were provisions meant to protect taxis, such as one that required drivers to pass a written test. To unlock the true potential of an on-demand transportation service, Uber needed to grow its supply of drivers without such obstacles. In February 2014, with the French government unwilling to relax the licensing requirements, Uber introduced ridesharing in France that allowed drivers without professional taxi licenses to pick up passengers using their own cars. Because Uber already had UberX in the country, a service that used licensed professional drivers, the company called this offering UberPop; regional general manager Pierre-Dimitri Gore-Coty chose the name with colleagues because it reminded him of the phrase *peer to peer.*

UberPop grew steadily in France until the summer of 2015, when taxi drivers across France protested, snarling highways, overturning Uber cars, and blocking access to Charles de Gaulle

Airport. The Interior Ministry, which regulates French taxis and enforces French laws, was sympathetic to the drivers and apparently to traditional taxi interests. Uber's Paris offices were raided and drivers were fined.[6] On June 29, 2015, authorities arrested Gore-Coty and Thibaud Simphal, Uber's general manager in France.[7] They spent the night in jail, and a few days later Uber shuttered the UberPop service, though it maintained its licensed-driver services in the country. After a 2016 trial, the executives and the company were found guilty of "misleading commercial practices" and ordered to pay a fine.[8]

And so it went in Italy, where a Milan judge outlawed UberPop in May of 2015, citing unfair competition;[9] in Sweden, where thirty drivers were convicted of operating an illegal taxi service, forcing the company to suspend UberPop there;[10] in Spain, where a judge banned Uber for a year, charging it with "unfair competition," and ordered Spanish internet providers to block access to the Uber app within the country; this came after a formal injunction from Spanish taxi companies and their powerful trade group, the Association Madrileña Del Taxi;[11] and in Germany, where judges ruling on claims by taxi trade groups found that Uber violated competition laws and should be required to use only professional licensed drivers.[12] Uber withdrew from the German cities Frankfurt, Hamburg, and Düsseldorf while maintaining its services using licensed drivers in Berlin and Munich.

The fight over Uber in each European country was revealing. On the one hand, it reflected Uber's clumsy aggressiveness, its proclivity to roar into cities with guns blazing and without cultivating allies in government, only to face an inevitable backlash later. "We made some mistakes," says Ryan Graves, Uber's head of operations. "We were a little more bull-in-a-china-shop than we needed to be."

On the other hand, Uber's expansion also measured the will of local governments to update antiquated transportation laws for a service that many of its own citizens desperately wanted. This was a litmus test for democracy itself, exposing whether regulators and legislators were more beholden to their own people or to powerful taxi interests and unions. The countries of continental Europe struggled with this test. They had encountered an innovative and arrogant new player out to disrupt a stagnant industry, and their impulse was to repel the upstart.

On the other side of the world, however, things were quite different. Asia's response to Uber's global ambitions—unlike Europe's—was primarily entrepreneurial. In fact, Travis Kalanick was about to come face to face with someone just as driven and aggressive as he was.

Back in the spring of 2012, news of the funding and imminent expansion of the British taxi-hailing service Hailo flooded the tech blogs. As we have seen, Hailo's premature announcement pushed Kalanick to quickly add vehicle choices to the Uber luxury-car app. Hailo was forced to retreat from the United States and return to its niche facilitating taxi rides in England and Ireland. In 2016, it was acquired by Daimler.

But Hailo's ill-fated expansion attempt in 2012 had another, even more profound effect on the progression of this story.

Halfway around the world, in the headquarters of the Hangzhou-based Chinese e-commerce giant Alibaba, a talented young salesman named Cheng Wei read about the coming Hailo-Uber showdown in the tech blogs and started plotting to take advantage.

Cheng Wei was born in Jiangxi Province, a landlocked region in eastern China famous for being the cradle of Mao Zedong's Communist revolution. His father was a civil servant, his mother a mathematics teacher. He excelled at math in high school but says that during his college entrance exams, he neglected to turn over the last page of the test, leaving three questions blank.

He got into the Beijing University of Chemical Technology, less prestigious than the upper-echelon schools. Cheng had planned to major in information technology but was instead assigned by his university to business management. He worked during his senior year, as Chinese students often do, selling life insurance. He didn't sell a single policy—not even to his teachers, one of whom told him that "even my dog has insurance," Cheng says. At a job fair, he applied to be a manager's assistant at a business that billed itself as a "famous Chinese health-care company." But when he showed up for work in Shanghai, luggage in hand, he discovered it was actually a chain of foot-massage parlors.

In 2005, out of school at twenty-two, he got an entry-level job at Alibaba by showing up at the front desk of its Shanghai office and asking for work. He landed in sales and earned 1,500 yuan, or $225, a month. "I am very thankful toward Alibaba," Cheng says. "Because someone stepped forward, didn't shoo me away, and said, 'Young people like you are what we want.'"

Despite his earlier insurance fiasco, Cheng proved to be good at selling online ads to merchants. He moved up the ranks and eventually reported to an outspoken executive named Wang Gang. When he first met Cheng, Wang says, the young man's sales numbers were strong, but his real talent was in emceeing customer events.

In 2011, Wang, unhappy about being passed over for a promotion, gathered Cheng and other underlings to brainstorm ideas for startups. They tossed around business models for companies in education, restaurant reviews, even interior decorating. In early 2012, they started tracking a smartphone app called Momo that allowed people to identify the locations of other users on an online map. The notion of tracking attractive females on their phones got them interested in a smartphone's GPS capabilities. That's when Cheng Wei read about Hailo's coming expansion to the United States.

To Cheng Wei, the Hailo news was a wakeup call. The United States and the United Kingdom were battling to drag the taxi industry into the smartphone era. Cheng knew that China had a massive taxi market that was regulated, stubbornly analog, and highly fragmented, with dozens of taxi companies in every major city.

He left Alibaba in 2012 and named his new taxi-hailing application Didi Dache, or "honk honk call a taxi." His boss Wang Gang also left Alibaba and became Cheng's primary financial supporter, investing 800,000 yuan, or about $100,000, in the startup. (Estimated value of Gang's stake at the end of 2016: around $1 billion.)

Cheng and several ex-Alibaba colleagues initially set up shop in a warehouse in northern Beijing, a shabby hundred-square-meter space with a single conference room. It turned out that their idea for a Hailo-like app for Chinese taxis was not really that novel. At least thirty other groups of entrepreneurs either saw the Hailo announcement or sensed the frisson of excitement in the air around electronic taxi hailing and were developing similar startups inside China at pretty much the exact same time.

Electronic cab hailing caught on especially fast in China, with its crowded subways, congested highways, and chronic smog that made it unpleasant to walk or bike. But at first, the business didn't seem like a particularly good one. Competition was fierce, and the taxi-hailing startups had to pay cabbies to help defray the cost of owning cell phones. The Chinese government, worried about any increase in transportation costs, prohibited the startups from collecting commissions on fares and in some cities even ruled the apps illegal—though drivers used them anyway, often carrying spare phones to show to inspectors if they were pulled over. Cheng says he dispatched two of his first ten employees to launch the service in Shenzhen, home to Foxconn's iPhone factories, because he thought of all of China's cities, Shenzhen had the most liberal regulatory attitude. Didi was promptly halted by local authorities.

Cheng is cherubic and bespectacled—he wouldn't look out of place in a video-game arcade at two a.m. He's recalling all of this from his spacious office, furnished with business books and a desktop goldfish tank, in the north part of Beijing.[13] On clear days, which are rare in the city, he can see the mountains in the northwest where the Chinese strengthened the Great Wall in the fifteenth century to protect against invasion from the Mongols. Considering everything that was about to happen, this seems apt.

All the early Chinese ridesharing startups lost money, and the ones that arrived late to the market or tried to replicate Uber's original strategy of starting with the more expensive but rarer black cars were critically handicapped. But Didi was scrappier than most of its rivals. When Yaoyao Taxi, a rival backed by Silicon Valley's Sequoia Capital, won an exclusive contract to recruit drivers at the Beijing airport, Didi employees descended on the city's biggest railway station to promote their app. Instead of imitating competitors and giving away smartphones to drivers, an expensive proposition for a capital-strapped startup, they focused on providing their free app to younger drivers who already had phones and were likely to spread the word about Didi.

During an epic Beijing snowstorm late in 2012, when it was impossible to hail a cab on the street, residents turned to the app, and the company surpassed one thousand orders in a single day for the first time. That got the attention of a Beijing venture capital firm, which put in $2 million, valuing Didi at $10 million. "If it didn't snow that year, maybe Didi wouldn't be here today," Cheng says.

Then, in April 2013, one of the startups established an early advantage—and it wasn't Didi. Kuaidi Dache ("speedy taxi"), based in the eastern city of Hangzhou, raised a round of funding from Cheng Wei's old employer, Alibaba.[14]

Dominant market share on the internet in China is often established by the startup with the strongest link to one of the Big Three—the entertainment portal Tencent, the search company Baidu, and the e-commerce giant Alibaba. These companies control the online landscape in the world's most populous country and are able to send torrents of traffic to their partners. Didi was catching on among techies in the cities of Beijing and Guangzhou, but to survive, Cheng Wei realized, he would need to forge an alliance. A few weeks after the Kuaidi-Alibaba deal, he raised $15 million from Alibaba's archrival Tencent in a round that valued the still-tiny company at $60 million.

With the backing of two rival Chinese internet giants, Didi and Kuaidi trained their sights on each other. During one notoriously difficult week, reverently known at Didi as "Seven Days, Seven Nights," both companies had intermittent technical problems, sending drivers and riders scurrying from one service to the other and back again. Cheng says engineers were holed up in Didi's cramped offices for so long and worked so hard to resolve their issues that one employee had to have his contact lenses surgically removed.

Finally, Cheng called Pony Ma, the founder and CEO of Tencent, for help. Ma agreed to lend him fifty engineers and a thousand servers, and he invited Didi's team to temporarily work out of Tencent's more comfortable offices.

But Didi wasn't making any money, and Cheng needed to raise capital. He visited the United States for the first time in November 2013, only to be rejected by multiple investors. "We had burned a lot of money," he says. "Investors were like, 'Whoa.'" There was another snowstorm that Thanksgiving in New York City but this one was less serendipitous. Cheng Wei says his Uber got caught in the storm on the way to the airport and he ended up missing his flight. "I was very depressed after I came back to China," he says.

In early 2014, everything changed. Over the 2014 Chinese New Year, Tencent had a huge hit with a mobile app called Red Envelope, which allowed users to send friends and family members small financial gifts for the holiday, an ancient Chinese custom.

Suddenly Alibaba and Tencent awoke to a new battleground in their long-standing war—mobile payments. Managing the primary online wallet for smartphone users in China could be a powerful strategic position. So both companies scrambled to establish their payment apps. Didi and Kuaidi were turned into proxies in this mad dash. Didi was integrated into the payment feature of Tencent's hugely popular chat app, WeChat, while Kuaidi allowed customers to pay using Alibaba's mobile-payment subsidiary,

Alipay. Both Alibaba and Tencent started funneling cash into their affiliated taxi apps, which then offered generous guaranteed payments for drivers and discounts for riders as a way to lure users to the dueling mobile-payment services.

Between 2009 and 2014, Uber had expanded in the United States with extraordinary speed, carried along by a great wave of booming smartphone usage. But in China, largely thanks to the furious competition between the tech giants and their desire to push their messaging and mobile-payments products, the wave nearly resembled a destructive tsunami. By doling out generous subsidies in 2014, Didi was burning through a hundred thousand dollars *every day* in the proxy war with Kuaidi, according to one of its investors. That year it raised $800 million in two separate rounds of funding from Tencent and the Russian venture capital firm DST Global, among other investors, while Kuaidi raised nearly as much from Alibaba, the Japanese tech conglomerate Softbank, and the private equity firm Tiger Global.[15] Cheng Wei was proving himself a clever and adaptable CEO, but at this rate, the battle with Kuaidi was going to be financially ruinous for everyone involved.

Didi's and Kuaidi's investors eventually realized the folly of their mounting rivalry. With Travis Kalanick starting to eye China as Uber's next big opportunity, they urged an armistice between the two startups and their corporate backers.

The savvy Russian venture capitalist Yuri Milner of DST helped to broker the merger, shuttling between Alibaba and Tencent's headquarters. Thanks to Cheng Wei's scrappiness and the strength of Didi's integration with WeChat, Didi now had more ride volume and ended up controlling 60 percent of the combined company. Cheng Wei "was essentially as aggressive as Travis," says a Didi investor. "He was like a perfect match."

Uber had been laboring quietly in China for two years. After completing the Series C funding over the summer of 2013, Travis

Kalanick had traveled to Asia on a celebratory trip with executives from TPG Capital. Before he left, he asked a group of colleagues to meet him in Beijing. Austin Geidt; Allen Penn, the former Chicago manager; Sam Gellman, an Uber executive living in Asia; and Corey Owens, the head of public policy, met Kalanick in Beijing, where they spent two weeks working from several "shoddy apartments in some corner of Beijing that I have yet to ever find again," Penn recalls.

For U.S. internet companies, expanding into China had long been considered a suicide mission. Google, eBay, Amazon, Facebook, and Twitter had all tried and failed to crack the world's second-largest economy, repelled by government censorship, the native advantages of the Big Three, or both. Kalanick, characteristically, was undeterred. With colleagues, he put together a list of all the reasons he believed they were different than their tech predecessors and concluded that they were creative and patient enough to succeed. He was a problem solver and this was the ultimate problem—one no other tech entrepreneur had overcome.

That week in 2013, Uber's away team fanned out in Beijing, experimenting with the local taxi-hailing apps, meeting with lawyers and regulators, and learning everything they could about the rules and realities of the country's taxi industry. Kalanick met with a number of startup CEOs, including a young Cheng Wei, who was only six months into running Didi at the time and who impressed Uber's boss. "Travis had met him even before I started," says Emil Michael. "He told me that among all the ridesharing founders, Cheng Wei was special. He was just a massive cut above anyone else in the industry."

The Uber execs all discovered the challenges of getting around in Beijing. Allen Penn recalls leaving himself ninety minutes to make it to a meeting across the city and spending half an hour trying unsuccessfully to hail a cab; eventually he went back inside, frustrated, and called into his meeting on Skype.

Everyone they met that week urged the Uber execs to proceed

cautiously in China and forge a joint venture with a local player. "Take your time. Do not rush this. American companies will get this wrong" was the gist of the advice, says Austin Geidt.

But as he was demonstrating in Europe, Kalanick was not in the habit of moving slowly. One day on the trip, he retrieved a few spare iPhones from his luggage and slipped in local SIM cards. He called and woke an Uber engineer in San Francisco and asked him to hack together a version of the Uber driver app for Beijing. Then Allen Penn and Patti Li, a TPG investor who spoke fluent Mandarin, found some willing drivers, and that night the visiting execs became the first Uber riders in China. "From a GPS standpoint it was a mess," Geidt recalls, since many Google services were blocked in China and Google Maps was an unreliable guide.

It would take another year for Kalanick and his execs to grow comfortable with the idea of launching in China. In early 2014, Uber rolled out its luxury black-car service in Shanghai, Beijing, Guangzhou, and Shenzhen. It charged in U.S. dollars only, at first, to position it as a tool for tourists and expats. Careful not to provoke the Chinese government, it intentionally didn't solicit any media attention. "We didn't want to come here loudly," Geidt says.

Uber muddled along quietly in China for a year as Didi and Kuaidi dueled under the aegises of Alibaba and Tencent. Then, in the fall of 2014, buoyed by Uber's success elsewhere in the world, Kalanick and his execs decided to introduce ridesharing in China. "This is where true entrepreneurs show up," says Emil Michael. "We thought, What's the worst that could happen? We're not an incumbent, so let's take the bet."

In October 2014, in Guangzhou, Shenzhen, Hangzhou, and Chengdu, Uber introduced the equivalent of UberX, which it called the People's Uber, allowing any drivers who cleared background checks to pick up passengers with their own vehicles. At the same time, it found a strategic partner who could offer money, valuable technology, and political connections with the Chinese government—the one member of the Big Three that had missed the expensive taxi-app wars and was late to the mobile-payments landgrab. In December, Baidu announced it was making an investment in Uber and that Uber would now run on the more dependable Baidu Maps in China.[16]

The strategy seemed to work, at first. With Didi and Kuaidi consumed with their merger, Uber started gaining ground on the strength of ridesharing and clawed its way to what it estimated was 30 percent of the Chinese market for on-demand transportation apps.

As usual, there was drama. Taxi drivers went on strike in half a dozen cities, including Changchun, Nanjing, and Chengdu.[17] The police raided Uber offices in Guangzhou and Chongqing.[18] In January 2015, the country's Ministry of Transport ruled that private car owners were not allowed to use ride-hailing apps for profit. But strangely, Uber and its rivals were allowed to continue to operate. The Chinese government showed little appetite for a total crackdown. It wasn't going to exterminate a service that promised to address the country's considerable transportation woes.

Uber now had leverage, and Travis Kalanick was going to try to use it. On a trip to Beijing, Kalanick and Emil Michael visited the Beijing offices of the newly merged and renamed Didi Kuaidi ("honk-honk speedy") and met with a group of executives that included Cheng Wei and his new chief operating officer, Jean Liu, a former managing director at Goldman Sachs. The meeting started well, by all accounts. Cheng Wei greeted Kalanick with the words "You are my inspiration," but after that the mood became tense.[19] Emil Michael remembers what he thought might have been a form of psychological warfare: "They served us maybe the worst lunch I've ever eaten," he says. "We were all just poking at our food, wondering, Is this some kind of competitive tactic?" (It wasn't. Jean Liu later apologized to Michael for the food.)

At one point during the meeting, Cheng walked over to a whiteboard and drew two lines. Uber's line started in 2010 and went up sharply and to the right, depicting its rapidly rising ride volume since its inception. Didi's started two years later, in 2012, but had a steeper curve and intersected Uber's line. Cheng said Didi would one day overtake Uber because China's market was so much larger and many of its cities restricted the use and ownership of private cars to manage traffic and pollution. "Travis just smiled," Cheng recalls.

According to Cheng Wei, Uber's CEO wanted to invest in Didi Kuaidi. He asked for a 40 percent ownership stake in the company and, in return, promised to cede China to Didi. In a speech, Cheng Wei later said that Kalanick threatened an "embarrassing defeat" for Didi in China if he rejected the offer. "We could tell from the way they looked at us that they thought of us as just another local taxi app from Sichuan," Cheng Wei said. "Foreign companies see China as a territory to be conquered."[20]

Jean Liu, a native of Beijing who spoke English fluently and who was Didi's primary liaison to the global business community, said that Kalanick came off as a bully. "Imagine someone coming to your office saying, 'Give me this much stake of your company, otherwise I will fight you,'" she says. Uber later disputed Didi's account and characterized the meeting as "super friendly."[21]

Didi execs rejected the proposal and soon introduced their own version of ridesharing in China, as well as carpooling options and commuter buses. Didi would prove a powerful incumbent, capable of raising billions of dollars of venture capital and going head to head with Uber to entice Chinese drivers and riders with discounts. It was going to be a global mega-unicorn death match—over the largest transportation market in the world.

On the afternoon of June 3, 2015, Uber invited local journalists to its Market Street headquarters in San Francisco to mark a momentous occasion: the fifth anniversary of the company opening its app to drivers and riders. Garrett Camp kicked off the event marveling that "a crazy idea" had turned into a globe-spanning juggernaut. Austin

Geidt and Ryan Graves each recalled when Uber was composed of only a few employees sitting around a cramped conference-room table in borrowed offices near the Transamerica Pyramid.

Then Travis Kalanick took the podium, looking nervous and emotional, with his parents sitting in the first row. Over the next twenty minutes, speaking awkwardly from a teleprompter, he acknowledged the aggressiveness that had made Uber such a polarizing company over the last half a decade. "I realize that I can come off as a somewhat fierce advocate for Uber," he said. "I also realize that some have used a different a-word to describe me."

Kalanick then made a political case for the company that he had never before so artfully articulated. Uber, he said, brings new transportation options to low-income neighborhoods that aren't served well by yellow taxis. It creates flexible jobs for the unemployed, for immigrants, and for students looking to finance their education. By combining riders in the same car via the carpooling service UberPool, the company was further lowering the price of for-hire vehicles and potentially taking cars off the road and reducing CO_2 emissions. "That's what we believe is the real game changer and those are the things we'll be working on in years to come," he said.

The scripted event attempted to showcase a more introspective and optimistic Kalanick. It bore the fingerprints of Uber's increasingly professional communications team led by David Plouffe, who would soon be replaced by Rachel Whetstone, the former head of communications and public policy at Google. But the message had another target besides the journalists in attendance—regulators and legislators in Europe and, particularly, on the East Coast of the United States, where a new wave of opposition to Uber was gathering at that very moment in New York City, the biggest taxi market in the country.

In spite of the victories that had allowed the taxi apps to establish their businesses there, New York City under Mayor Bill de Blasio was largely hostile to Uber, just as it had been to Airbnb. In early 2015, the city and Uber battled over legislation to cap surge pricing[22] and over whether the company should turn over its trip data to the Taxi and Limousine Commission.[23] That May, the TLC considered severe restrictions that would, among other things, give it the authority to review any changes to the Uber application.

There was mistrust on both sides. The city accused Uber of being intransigent and refusing to play by the rules. Uber and its surrogates alleged that de Blasio was listening to his friends in the yellow-taxi industry, which had donated significantly to his mayoral campaign.[24] Both accusations were likely true. It was also true that the ground was shifting dangerously under the taxi fleets in New York. The value of a taxi medallion had peaked at more than $1.2 million in 2013, making them affordable only to fleet owners with access to bank loans or multiple drivers who could pool their money. They were selling for less than half of that by 2016.[25] David Yassky, the chairman of the TLC between 2010 and 2014 (and a Lyft advisor), told me he felt the new restrictions were "pure protectionism."

Uber organized protests with riders and drivers at city hall, urging the city to abandon the new rules. On June 18, 2015, the TLC seemingly backed off. There was a brief respite from conflict, with both sides praising each other in the press.[26] But then, twenty-four hours later, Meera Joshi, Yassky's successor as TLC commissioner, called Michael Allegretti, a public-policy manager at Uber, and told him that a law was about to be proposed in the city council to cap the number of new private-hire licenses granted to Uber, Lyft, and other app companies, pending the outcome of a study on congestion in lower Manhattan. "There's nothing you can do to stop that one," Joshi told Allegretti. "The votes are there."

The bill was proposed the next day, and it was as bad as Uber could have imagined. Under the legislation, Uber and Lyft could grow their supply of drivers by only 1 percent a month while the congestion study was conducted, which could take a year or more.[27] Capping Uber's supply of drivers was paramount to freezing its growth and could provide a road map for opposition forces in places like London and Mexico City. The city council planned to vote on the bill in twenty-one days.

This, Uber execs reasoned, was a hit job. The downtown streets had indeed slowed to a crawl, but that was the result of many factors, including the proliferation of bike lanes, reduced speed limits, a booming economy, an increase in e-commerce delivery trucks, new construction, and on and on. Uber would have to fight this the old-fashioned way, in the trenches.

Summer 2015's three-week battle for New York's streets offered a fresh look at what was now a well-funded, well-organized, and ruthless political operation. The campaign hit the progressive Democrat de Blasio from the left on jobs and equal access to transportation in the outer boroughs, issues that mattered to the constituencies that had gotten him elected in the first place. Uber's campaign included mailings, robo-calls, a rally of drivers in Queens, and a pair of brutally effective television ads that ran widely that month in the New York area.

The ads featured a series of African American and Latino Uber drivers talking to an off-camera interviewer, crediting Uber for giving them work and levying some indirect charges against the yellow-taxi industry and the mayor.

"People have access to an Uber in places where they never thought they would be able to be picked up."

"This is New York. We live in five boroughs!"

"The mayor is giving in to the taxi industry."

"He should know the struggles that most New Yorkers go through. Embrace the fact that people want to go to work!"

"When the mayor came to town he promised to provide jobs."

There was little subtlety to the parade of minority faces; Uber was beating de Blasio over the head with an implicit race-based appeal. David Plouffe, now an Uber board member and adviser, hammered away with that message, appearing on TV, meeting with newspaper editorial boards, and holding a press conference with African American community leaders at Sylvia's, an iconic soul food restaurant in Harlem.

And in a clever display of political jujitsu, the company added a feature to its app called De Blasio's Uber, which showed New York's two million Uber users a dystopian future where the wait to be picked up was a horrifying twenty-five minutes. Kaitlin Durkosh, a junior member of the communications team, had come up with the idea, which was catnip to media covering the fracas. "This is what Uber will look like in NYC if Mayor de Blasio's Uber cap bill passes," the app informed its users. It then asked them to e-mail city hall and make their displeasure known.

Lyft lobbyists were also active in the fight, though quieter. Lyft's representatives met with city council members that month to press their case and highlight the congestion-relieving benefits of their own carpooling service, Lyft Line. Their most effective argument was that freezing the number of private drivers would only entrench Uber's advantage. That resonated, says one Lyft exec. "Whoever we were meeting with said, 'Look, I hate those Uber guys. They are the worst. Help me figure out how to make only Lyft legal.' "

The end for de Blasio in the Uber fight was swift and humiliating. A day before the vote, New York governor Andrew Cuomo, who had sparred openly with de Blasio on issues such as charter schools and upper-income tax increases, announced his opposition to the bill and suggested the state might get involved. "I don't think that government should be in the business of trying to restrict job growth," the governor said, twisting the knife in his political rival.[28]

The next day at noon, Allegretti, the Uber public affairs exec, got a phone call. The mayor's office wanted to talk. He went to city hall, on 250 Broadway, with East Coast regional manager Rachel Holt, NYC manager Josh Mohrer, and Justin Kintz, Uber's head of public affairs, and met with de Blasio's political director and deputy mayor, among other officials. The conversation was brief—the mayor was going to drop the cap provision pending the outcome of the congestion study. (The study would show that tourism, increased construction, and deliveries were mostly responsible for lower Manhattan gridlock.)[29]

It was another victory for Uber, total and sweeping. In less than a month, the company had put together an unlikely coalition of affluent riders and minority drivers from all five boroughs. It had demonstrated that Travis's Law still reigned in the United States, that as long as people loved Uber they would fight for it, and that the taxi industry had few friends. These lessons would prove handy in other American cities, like Las Vegas, Austin, Portland, Miami, and wherever battles over ridesharing were being waged. Uber would win many of these fights, lose a few, and demonstrate that it still had capital, political connections, a great many thousands of fervent customers, and the grand arc of history itself on its side.

In the fall of 2015, Travis Kalanick treated his five thousand employees to a lavish, four-day, all-expenses-paid retreat in Las Vegas. It was part all-hands meeting, part gaudy celebration of... well, it wasn't quite clear of what, or even that the company needed a reason to celebrate. But Uber's PR team recognized that the trip was likely to play poorly in the press and among Uber's drivers, so it impressed upon attendees that they were not allowed to post anything about it on social media. The secrecy was so extreme that Uber created a special logo for the occasion, two *X*s inside a box, so that bystanders couldn't identify the company. Nevertheless, the *Daily Mail,* a British newspaper, ran an account of the event, and several current and former Uber employees later shared with me their recollections.[30]

Employees were put up, two in a room, in five hotels on the Vegas Strip. During the day there were seminars on topics such as supply growth and business development as well as optional philanthropic expeditions to local food banks. In the afternoons, employees luxuriated and drank standing in the hotel pools in the 90-degree desert heat. At night there were dinners and talks, including one Q-and-A session with Kalanick and media entrepreneur Arianna Huffington, a future Uber board member, and another with Uber investors Bill Gurley and Shervin Pishevar. Afterward there were dance parties and more poolside revelry, which apparently wasn't for everyone. "That's when I realized how millennial the company was," said one employee who quit a few months later, worn out by the grinding internal pace. "I'm thirty-five. I don't want to stay out until three in the morning. I felt ancient."

Tuesday night was the main event and an indication that Kalanick was endeavoring to coax his upstart into adulthood. Employees filled the amphitheater at the Planet Hollywood Resort and Casino, where Kalanick commanded the stage for two and a half hours, wearing a white laboratory coat and unveiling the company's newly conceived cultural values.

Cultural values can be a rudder for large companies, a way to align thousands of far-flung employees and guide the hiring of new workers with a set of rigorously defined ideals. Airbnb had formulated its six values back in 2012 ("Be a host" and so forth), and they had helped to shape its conciliatory manner of dealing with unanticipated crises and regulatory turmoil. Uber had skipped this step earlier in its history, which was apparent in its more slapdash and aggressive approach to unanticipated obstacles.

Kalanick called his new values a "philosophy of work" and said he had deliberated over them for hundreds of hours with colleagues, including his chief product officer, Jeff Holden. Holden was a former Amazon executive and a disciple of Jeff Bezos, and it showed; many of Uber's principles were comparable to those of the widely admired technology giant, and, like Amazon, Uber had fourteen of them. Onstage at Planet Hollywood, Kalanick discussed each (the parenthetical descriptions are mine):

Customer obsession (Start with what is best for the customer.)
Make magic (Seek breakthroughs that will stand the test of time.)
Big bold bets (Take risks and plant seeds that are five to ten years out.)
Inside out (Find the gap between popular perception and reality.)
Champion's mind-set (Put everything you have on the field to overcome adversity and get Uber over the finish line.)
Optimistic leadership (Be inspiring.)
Superpumped (Ryan Graves's original Twitter proclamation after Kalanick replaced him as CEO; the world is a puzzle to be solved with enthusiasm.)
Be an owner, not a renter (Revolutions are won by true believers.)
Meritocracy and toe-stepping (The best idea always wins. Don't sacrifice truth for social cohesion and don't hesitate to challenge the boss.)
Let builders build (People must be empowered to build things.)
Always be hustlin' (Get more done with less, working longer, harder, and smarter, not just two out of three.)
Celebrate cities (Everything we do is to make cities better.)
Be yourself (Each of us should be authentic.)
Principled confrontation (Sometimes the world and its institutions need to change in order for the future to be ushered in.)

Kalanick showed several slides and a video for every value and ended each topic by calling on an executive to illustrate it with a story or observation. Here was an encore for some of the main figures in Uber's history; Ryan Graves, Austin Geidt, Rachel Holt, Allen Penn, and Holden himself all took their turns with the microphone, sharing their personal stories with the company.

Employees who hadn't sufficiently glugged the company Kool-Aid thought it was an overly long and self-indulgent display. Others called it their most formative experience at the company. "It was one of the most moving moments I had at Uber," says Austin Geidt. "You could see how big we were, how many different countries were represented, and all the different types of people. It was awesome."

Afterward the employees lined up for buses to take them to another nightclub, where the DJs Kygo and David Guetta performed. And the next night, the lucky Uber employees were treated to a private performance by an Uber investor, the megastar Beyoncé.

A few months later, on February 1, 2016, a few hundred Uber drivers gathered in front of the company's offices in Long Island City to protest the latest round of UberX fare cuts. Uber had recently reduced its fares by 15 percent in many cities, part of its annual effort to stimulate demand during the winter slowdown and increase the frequency of rides (and, no doubt, to apply further financial pressure to its domestic rival Lyft). NO ONE WINS THE RACE TO THE BOTTOM! read one sign. GIVE US THE RATE BACK. SHAME ON UBER! read another.

The drivers that day in Queens were angry about what now seemed like a sub-minimum wage, about Uber's ever-increasing commission, and about the need to work longer hours to eke out a living. Uber promised to pay drivers a minimum hourly rate if their earnings fell below a certain level, but the drivers said that the company found a myriad of ways to disqualify them from the guaranteed wage. They also complained that Uber, unlike Lyft, had consistently refused to allow passengers to give them tips over the app.

"Nobody in America wants to work more and earn less," said Mohsim, a Pakistani driver wearing a silver badge—a symbol, he said, of the Ottoman Empire. "It's modern-day slavery."

Another protester, Angel, forty, said he expected his pay to drop 20 percent from last year. "What if I took ten thousand dollars out of your annual salary?" he said. "It's double the work for less money." Angel said that Uber had also drenched the city in billboards and bus ads soliciting new drivers, which had oversaturated the market and made it more difficult to get rides.

Uber insisted that the fare cuts were going to prove beneficial to drivers and promised to roll them back in cities where they didn't result in more rides and better earnings. But those promises, at least to the vocal subset of drivers in Queens that day, seemed empty. They felt disempowered, overworked, and even nostalgic for the government-set fares of the yellow-cab industry.

If the drivers had unrealistic expectations about Uber, that was at least partly because Uber itself had helped create them. A central premise of the company was that electronic hailing liberated drivers from the tyranny of taxi-fleet owners and mandatory twelve-hour shifts. "We give riders high-fives, but we give drivers hugs" was an oft-repeated phrase from Kalanick. And in a 2014 blog post, the company claimed that drivers in New York City earned $90,766 a year, while drivers in San Francisco earned $74,191.[31] These were easy numbers for journalists to debunk, and surveys showed they were inflated, particularly when you factored in the expense of commercial insurance and car leases.[32]

Uber referred to its drivers as small-business owners and entrepreneurs. But as enterprising fleet owners had discovered years before, it was impossible to maintain a business of any size on Uber, which remorselessly discarded all middlemen in favor of a direct relationship between individual drivers and the company. Drivers weren't really small-business owners. They most closely resembled taxi drivers at the whim of a distant master who had the principal goal of any corporate concern — to build as big a business as possible.

In this sense, Uber was part of an American business trend that had been playing out for decades: the categorization by profit-minded companies of workers as part-time contractors instead of employees. Since the early 1980s, companies had been circumventing minimum wage and other protections for employees, who filled out W-2s, by reclassifying them as contractors, who filled out 1099 tax forms. Labor groups and lawyers went to court repeatedly over the years to reclaim employment rights on behalf of truck drivers, waiters, housecleaners, exotic dancers, and even yellow-cab drivers. They mostly lost these cases, defeated by deep-pocketed corporations and, in 2011, a Supreme Court ruling that allowed companies to force their workers to sign arbitration clauses that prevented them from bringing class-action lawsuits.

Uber, Lyft, and the other companies in the so-called sharing economy gave plaintiff attorneys a high-profile opportunity to argue once again that workers were being stripped of protections. In 2013, Boston plaintiff lawyer Shannon Liss-Riordan brought such lawsuits against Uber and Lyft in the two states where she thought the law was most favorable, California and Massachusetts. She had previously brought similar, largely unsuccessful cases against FedEx and several yellow-cab companies. Uber's claim that it was facilitating a whole new kind of internet-enabled, on-demand work bothered her. "The mere fact that there is flexibility does not mean that the people who are doing the jobs shouldn't get benefits and the protections of employment," she says. "That is the reason we have these laws."

Both Uber and Lyft tenaciously fought against the cases, arguing that the great majority of their drivers didn't actually consider themselves full-time chauffeurs and wanted to remain independent and free to take other work.

The cases against Uber and Lyft drew widespread media attention and produced an unrealistic expectation that they might somehow change the nature of the sharing economy and undermine Uber's business model. (This was unlikely, as class-action lawsuits do not change the law.) The perception deepened when Liss-Riordan scored impressive victories in March 2015: judges in both cases said they could proceed to jury trials.

But a year later, the Ninth Circuit Court of Appeals agreed to hear Uber's argument that the lawsuit had been improperly certified as a class action and violated the drivers' arbitration agreements. Liss-Riordan, who had lost many such appeals on this issue, knew there was trouble ahead. Instead of proceeding to trial, she leveraged her victories into a settlement. Uber agreed to pay as much as $100 million to a group of tens of thousands of drivers and to institute new policies, such as giving drivers explanations if they violated company rules and got kicked off the app and creating an appeals process for those decisions. But Uber and Lyft drivers were going to remain contractors.

"Drivers value their independence—the freedom to push a button rather than punch a clock, to use Uber and Lyft simultaneously, to drive most of the week or for just a few hours," wrote Kalanick in a blog post titled "Growing and Growing Up" that announced the settlement. He conceded that the company hadn't "always done a good job working with drivers," but reiterated that Uber presented "a new way of working: it's about people having the freedom to start and stop work when they want, at the push of a button."

In August 2016, the entire settlement was struck down by a federal judge, who ruled that the payout to drivers was inadequate. It looked increasingly unlikely that the question of whether Uber drivers were being treated fairly or if they should be considered employees was going to be adjudicated in the courts.

But it did come into clearer focus for me at the Uber Partner Support Center on Chicago's economically distressed South Side, across the street from a fenced-off, abandoned mall. A former Marine named Robert Davis helped to run the center and spent his day onboarding new Uber drivers and explaining how to use apps and smartphones to non-tech-savvy people. He had grown up in the nearby Auburn Gresham neighborhood and said Uber brought jobs and transportation to a community that historically had not had much of either.

Over the past year, he said, he had signed up single mothers, college students who needed extra cash, and widows looking for something to do with their time. Driving for Uber was sometimes a primary job but often a secondary one, supporting the other things they were trying to do with their lives. (In fact, the company says, 60 percent of Uber drivers are on the road ten hours or fewer a week.) "I can see both sides of everything," Davis told me. "I don't know why this is controversial. To me it seems like Uber is very targeted. It helps people who need extra income."

* * *

Meanwhile, in China, the mega-unicorns battled. It had once seemed that Uber enjoyed insurmountable advantages. It had a better app, powered by more stable technology. At the beginning of 2015, investors were valuing it at $42 billion, about ten times Didi's valuation. "At that time we felt like the People's Liberation Army, with basic rifles, and we were bombed by airplanes and missiles," says Cheng Wei. "They had some really advanced weapons."

Cheng was a student of military history and was particularly interested in heroic conflicts like the battle of Songshan during World War II, when Chinese nationalist troops tunneled under a mountain to surround the invading Japanese army. Uber execs met in San Francisco in what they called the war room, and Cheng held morning meetings with the Wolf Totem, his senior staff. The name, based on a popular novel set during the Cultural Revolution about urban students sent to live in Inner Mongolia, connotes aggression. The Wolf Totem studied Didi's daily results and adjusted the subsidies given to drivers and riders. Cheng would regularly warn employees, "If we fail, we will die."

In May 2015, Cheng went on the offensive. Didi said it would give away one billion yuan in rides. Uber matched it. Cheng and his advisers searched for ways to fight the American company on its home turf. Uber, they reasoned, was like an octopus — its tentacles were everywhere in the world, but its mantle was in the United States. Wang Gang, the early investor and board member, suggested at a meeting that Didi "stab Uber right in its belly."

Gang says Didi contemplated expanding into the States. Instead, in September 2015, it invested $100 million in Lyft. Then it established an anti-Uber ridesharing confederacy with Lyft and the regional ridesharing startups Ola, in India, and Grab Taxi, in Southeast Asia, all of them agreeing to share technology and integrate with one another's apps. According to Gang, it was less about undermining Uber than about gaining negotiating leverage. "The purpose of them grabbing a lock of our hair and us grabbing their beard isn't really to kill the other person," he says. "Everyone is just trying to win a right to negotiate in the future."

At the peak of hostilities, Didi and Uber were each burning through more than a billion dollars a year in China, giving unprofitable subsidies to drivers and riders. As Kalanick had expected, the ridesharing business in China was massive. Six of Uber's top ten cities by ride volume were in the country. Subsidizing rides at that scale made each company desperate for new capital. Uber raised over $4 billion in 2016, including a controversial $3.5 billion round from an unlikely investor—Saudi Arabia's Public Investment Fund. It pushed off an initial public offering and funneled capital to its Uber China subsidiary.

Didi Kuaidi, now renamed Didi Chuxing ("honk-honk commute") went toe to toe with its U.S. rival, raising $7 billion in 2016 and swelling its ranks to over five thousand employees, about a quarter of them working out of a collection of prefabricated five-story buildings on the periphery of Zhongguancun, Beijing's technology district. That summer, Didi claimed an 85 percent market share in China and operated in four hundred Chinese cities, while Uber was active in only one hundred. Uber's large institutional investors were worried and began pressing Kalanick to negotiate a truce.[33]

Cheng says the initial call for peace came from Uber; Emil Michael from Uber contends that the Saudi money forced Didi to the table—the investment suggested there was simply no end to the capital Uber could tap. Regardless, both sides agreed it was time to stop the bloodletting and focus on guiding their companies to profitability and investing in coming technologies, like driverless cars. "It was like an arms race," Cheng says. "Uber was fund-raising; we were also fund-raising. But in my heart I knew our money needed to be put into a more valuable field. This was why we were able to join hands with Uber in the end."

Emil Michael and Jean Liu hammered out the deal terms in two weeks. Uber agreed to depart China and hand over its operations in the country to Didi; in return it got a 17 percent stake in its Chinese counterpart and a billion dollar investment from Didi; the companies also took observer seats on each other's boards.

Michael and Liu met Kalanick and Cheng at a hotel bar in Beijing to raise glasses of baijiu, a traditional Chinese spirit made from sorghum. Over drinks, the CEO's spoke of mutual respect for how hard both sides had competed. "We are the craziest companies of our times," Cheng says. "But deep in our heart we are logical. We know this revolution is a technology revolution, and we are just witnessing the very beginning."

By fighting hard and being every bit as dogged as Uber, in investor boardrooms and on the battlefield, the Chinese CEO had repelled a foreign invader, ended the global mega-unicorn death match on auspicious terms and secured for Didi Chuxing its rightful place among the upstarts. "Cheng Wei is a fierce competitor. He has a champion's mindset," Kalanick told me, referring to one of his prized cultural values.

It was the highest possible compliment the chronically combative Kalanick could pay to an archrival. But it disguised a subtler truth: he too had pulled off something remarkable. Uber had spent more than $2 billion on a losing effort in China. But its stake in Didi, plus the $1 billion investment, was now worth $7.2 billion, at least on paper. It was an impressive return on capital. And through his stake in Uber, investors in both companies told me, Kalanick now owned nearly as much of Didi as Cheng Wei, whose personal stake had been diluted by the ceaseless mergers and fundraising rounds.

Colleagues say it took a few months for Kalanick to come around and agree to a surrender in China. Previously he had known only one mode of operation—aggression. Now, like his company, Kalanick was maturing and yielding to the lessons of a dynamic age. Pragmatism beat missionary zeal. Selective partnerships trumped solo adventuring. A disorganized and decentralized taxi industry had been conquered, but fresh challenges, with new sets of dangerous competitors, always loomed on the horizon. It had been eight adrenaline-fueled years of nearly non-stop conflict. With the Didi deal, the CEO of the world's richest, most valuable, most scrutinized upstart freed himself and his colleagues to finally confront the future.

Epilogue

By the end of 2016, Airbnb and Uber no longer resembled gangly, adolescent startups. They had thousands of employees, offices around the world, and ranks of experienced executives. In many cities, they still faced serious regulatory obstacles, but they now had vast political arms to wage those fights while they carefully laid the groundwork for their inevitable blockbuster IPOs.

Yet the true measure of an entrepreneur is how well he or she can identify new opportunities, and so, not surprisingly, in the fall of 2016, both Brian Chesky and Travis Kalanick were ready to talk about the future.

In October, I visited Airbnb at its bustling headquarters on Brannon Street. As usual, the impressive three-story green wall in the atrium was being tended to by gardeners, perched atop a scissors lift. It was lunchtime, so employees milled about the company cafeteria, which had recently been moved to the ground floor, next to a new high-end Spanish restaurant. I met Chesky upstairs in a conference room, where I found him wearing a company T-shirt with the Belo logo in African colors and fiddling with the audio of a presentation he had been working on for months.

In every way, Airbnb was booming. In August, it had its best night ever, with 1.8 million people from nearly every country in the world staying in properties they had booked on the site. More than one million listings were now available via the instant-book feature, which required no e-mailing back and forth with hosts. That's roughly equal to the number of rooms operated by Marriott International, the largest hotel chain in the world.

Chesky was getting ready to put all that at risk by betting on an even more ambitious vision: that Airbnb could broker not only apartments and homes but unique experiences for travelers. In internal discussions, he had dubbed this effort Magical Trips, using one of the favorite words of his idol Walt Disney, whose biography had inspired him to leave Los Angeles nine years before. Now he was preparing to introduce the service, officially called Trips, at the annual Airbnb Open conference later that fall in Los Angeles.

The centerpiece of the initiative was a totally revamped Airbnb app and website, which presented a Homes category and, alongside it, new tabs for Experiences and Places. With Experiences, travelers would be able to buy unique excursions, like hunting for truffles in Florence and visiting literary landmarks in Havana. Local entrepreneurs and celebrities would create and conduct these tours themselves. While the average price of an outing would be around two hundred dollars, Chesky showed me a deluxe eight-hundred-dollar experience created by the former sumo wrestler Konishiki Yasokichi that included a visit to a sumo training session, an undoubtedly bounteous meal with the wrestler, and prime seats at one of his sumo tournaments.

The other tab was a variation on the same idea. In Places, which Chesky called "our version of a guidebook," hosts and local luminaries would recommend the best things to see and do in their communities. Ideally they would point travelers not to tedious tourist traps but to farmers' markets, local theater productions, their favorite restaurants, and nearby charitable endeavors. Chesky imagines eventually letting visitors book restaurant reservations, tickets, and various forms of transportation from within the app, with the company taking a commission.

Nestled within the new service was a powerful idea—that Airbnb could rescue tourists from a narrow set of crowded, artificial travel experiences and direct them toward a larger set of more authentic interactions in real communities. "We had the former mayor of Rome here," Chesky told me, "and he said that mass tourism is a problem. You just have way too many people going into the Coliseum, and these old monuments can't take them all."

Chesky thought Trips, with its curated tours and endorsements, could entice travelers to fan out in major cities and visit destinations they otherwise wouldn't. "Most people don't come to Airbnb and say, 'Hey, I'd love to vacation in Detroit,'" he said. "And yet we think there's a ton of interesting culture [there]. It'd actually be an amazing trip probably, and it's going to cost a lot less."

The vision of Airbnb, he contends, was always to establish special bonds between hosts and guests, or, as he puts it, to offer "a people-to-people kind of culture and diplomacy." Chesky felt that Trips would give travelers another way to meet locals—not just their hosts but entrepreneurs and artisans, the actual residents of the places they visited.

Of course, it was also good business to try to sell other goods and services to Airbnb's vibrant customer base. (In the old days, this was simply called "upselling.") But Chesky, as usual, saw all of it in soaring missionary terms. As we talked, he framed Trips with increasing grandiosity—as a way to spawn millions of new friendships among strangers every day, to enliven cities and boost the micro-economy of artisans and entrepreneurs, even to give humans meaningful new work after the robots have taken all the jobs.

"The good news is I don't think that we will magically live in the first era of human history where you suddenly just run around the planet not knowing what to do," he said. "I'm an optimist. I do think there are many things people do. But I think, to be kind of simple about it, that people will do what only people can do. So can only people drive cars? I don't know. But only people can host other people. Only people can provide care. If you want something handmade, only a person can do that."

He even cast Trips in epic terms for Airbnb itself, as a way to invigorate the increasingly professional eight-year-old company with the kind of entrepreneurial energy that it once had in the founders' apartment on Rausch Street. And he compared the significance of Trips for Airbnb to Amazon's move from selling only books to selling other products in the late 1990s and to Apple upending the cell phone industry with the first iPhone in 2007. "I want to do for travel what Apple did to the phone," he said.

Chesky's presentation on Trips lasted an hour. As usual, it wove in Amol Surve and the first Airbnb guests back in 2007. And it was larded with his usual extravagant overtones. But it was also, I had to admit, a seductive vision. Living in a city, rather than just visiting it, was exciting. And having dinner in Japan with a sumo wrestler sounded delicious.

A few weeks later, with Uber's battle in China finally over, I visited Travis Kalanick in his San Francisco headquarters for our last interview. Unlike Chesky, he wasn't pitching anything, though Uber was just as actively rethinking its future. But he fielded a last round of questions with his usual chippy demeanor.

There was plenty to talk about. Uber now had nearly ten thousand employees, half of them in San Francisco. The company was about to break ground on a new two-building campus in the Mission Bay area of San Francisco and had purchased the ninety-year-old Sears building across the bay in Oakland, anchoring a recovering uptown business district.

I started by asking Kalanick about China and the end of the mega-unicorn death match. He said he had agreed to sell his business there when it became clear the battle might go on indefinitely. "Look, we could keep going. Both sides could keep going. It was just time," he said. "The thing for us was that the ridesharing wars were going global. You had American tech money going into our Chinese competitor and you had Chinese sovereign wealth money going into our global competitors. So then it was like, we are going to have to partner. This just made a lot of sense."

But why, I wondered, had he staked so much on China when the record of American internet companies in that country was so poor?

"It was almost like a romantic notion," he said, thinking back. "We wanted to engage in these places and learn and see if we could do something interesting and beautiful. I didn't want to miss out on that learning, on that experience. And by the way, there's also an economic and business case for it too. Competing makes you strong because it means you're serving riders and drivers better. As an entrepreneur, you want to see if the way you've created is a way that can work. Sometimes it's really easy to go and acquire a competitor, but we resisted that."

How did he explain raising more than $10 billion over the past two years alone—an inordinate amount for any company, let alone a privately held startup?

"If you didn't do it, it would be a strategic disadvantage, especially when you're operating globally," he said, noting that rivals like Didi and Lyft had also exploited a frenetic capital environment to raise funds for their own war chests. "It's not my preference for how to build a company, but it's required when that money is available."

When would Uber get to profitability? "We've been profitable in a number of cities for years. We raised a lot of money to invest in our operations. We can stop investing [whenever we want] and get profitable," he said. I noted there were skeptics who still thought the economics of ridesharing companies were unsustainable, propped up only by venture capital.

"Then how were we profitable in the U.S. in February? And maybe March too, I don't know," he said. "How could both of those things be true at the same time?"

Then he referenced Lyft, which was reportedly subsidizing rides aggressively in several big U.S. cities, trying to take market share from Uber. "We don't just go jumping in the lake with them, but we have to engage in some way," Kalanick said. "When your competitors are giving rides away, for forty to fifty percent off on the weekdays, you've got to respond."

As for when Lyft too had to get to profitability, he said he couldn't wait. "All businesses have to be disciplined and run sustainably, which is my DNA as an entrepreneur. So maybe it's like, I want to go back to my happy place... That's my sweet spot as an entrepreneur."

I changed the topic. Was ridesharing dead in Europe?

Kalanick didn't think so. "Progress eventually finds a way," he said. "Especially when the difference between what is being considered and what exists is so, so massive and obvious. You know, look, we've had issues and frictions or we just haven't gotten going in a few places, like Japan, South Korea, or Germany. But does that mean that ridesharing is never going to be a thing in those places? No, of course it will happen. I was recently in Germany and the number one value you can bring to the party in Germany is patience."

Uber drivers had continued to plead their cases against the company, and in Seattle they had even won the right to form a union. Was Uber treating its drivers fairly?

Kalanick equivocated a little. He had abandoned the pretense that driver earnings went up when Uber's fares went down and settled for contending that they remained steady. But he still genuinely seemed to think of drivers as Uber's customers. "I'd say the bottom line is that we have to show by all measures that [driver] earnings are stable," he said. Uber "needs to find ways to take stress and anxiety out of the work that is possible on the platform."

Finally, I asked: What is the future of Uber? How much of what is possible have we seen?

Kalanick started by declaring that on "logarithmic squared time," Uber was only halfway to its goals. This was the math geek from Granada Hills High School talking and it sailed over my head. But then he offered this:

"The things that people are going to feel are still to come. The kind of impact this is going to have on our cities—ninety-five or ninety-eight percent of it is still yet to happen. What if I said there's going to be no traffic in any major city in the U.S. in five years?"

"That would be a lot to live up to," I said.

"It would be. I think that might happen. We're just at the beginning. But when you feel that, that's going to be a big deal."

"This will be because of carpooling services like UberPool and Lyft Line? Or driverless cars?" I asked. A few weeks before, the company had begun testing fourteen Ford Fusions tricked out with autonomous vehicle technology on the streets of Pittsburgh. It had also recently announced a partnership with Volvo to develop driverless-car technology and had acquired Otto, a San Francisco–based startup run by former Google engineers that was working on driverless trucks.[1]

"All these things that are going to happen, whether it's human-driven transportation, carpooling, commuting with carpooling, driverless cars," Kalanick said, "cars are coming off the road. They're going to be much more efficient and much safer. And they're going to take up a lot less space. Our cities are going to be given back to us. Our time is going to be given back to us. And there's going to be a very, very different world in terms of how we experience our cities. We are just getting started."

Both Travis Kalanick and Brian Chesky had made big promises: to eliminate traffic, improve the livability of our cities, and give people more time and more authentic experiences. If these promises are kept, the results might be well worth the mishaps and mistakes that occurred during their journeys; perhaps they'll even be worth the enormous price paid by the disrupted.

And if they can't meet their own lofty goals? Or if the intensity of competition pushes them further toward a ruthless, win-at-all-costs mentality? Then Uber and Airbnb risk validating the worst claims of their critics—that they used technology and clever business plans merely to replace one set of dominant companies with another, amassing a staggering amount of wealth in the process.

I'm more optimistic than that. I believe in the power and potential of the upstarts and have frequently admired their resourceful, adaptive CEOs. But it's up to us to hold them to their promises. They are the new architects of the twenty-first century, every bit as powerful as political leaders and now completely enmeshed in an establishment that they have, at times, bitterly fought.

Eight years ago, they were at the inauguration of Barack Obama to witness the dawn of a new era. They didn't really invent their companies' respective ideas back then, but they honed them nearly to perfection, and then through acumen, fortitude, and sheer force of will, they enlisted large communities of users and persuaded at least some governments to step aside. And now the upstarts have the opportunity to make an even more significant impact.

But first, winter was coming, and along with it the challenges and ample uncertainties presented by the inauguration of a new American president. From now on, they would always remember to pack warm coats.

Acknowledgments

Writing a book about one fast-growing and secretive technology juggernaut is challenging enough. Weaving together the stories of two such companies as the competitive ground shifts beneath their feet every day is downright daunting.

So I owe a tremendous debt of gratitude to the editors, colleagues, and family members who supported this book from conception to finish. My agent Pilar Queen was an invaluable adviser, skilled at dispensing both practical advice and anxiety-defusing reassurances. At Little, Brown, my editor John Parsley believed in this book from the beginning and was insightful and wise at every juncture. Additional thanks go to Little, Brown's CEO Michael Pietsch and to Reagan Arthur, Nicole Dewey, Tracy Williams, Michael Noon, Lauren Harms, and Gabriella Mongelli for escorting this book through the gestation process. Tracy Roe gave the manuscript a sterling copyedit. A big thanks also to Doug Young at Transworld Publishers for championing this book in the UK.

I'm grateful to everyone at Airbnb, Uber, and Lyft who saw value in a deep look at a momentous era in Silicon Valley history. At Uber, thanks go to Jill Hazelbaker, David Plouffe, Nairi Hourdajian, and Travis Kalanick and his executives. At Airbnb, I'm grateful to Kim Rubey, Maggie Carr, and Mojgan Khalili, as well as Brian Chesky, Joe Gebbia, Nathan Blecharczyk, Belinda Johnson, and their team. At Lyft, Brandon McCormick had infinite patience for my inquiries, and John Zimmer and Logan Green were generous with their time and recollections.

In Silicon Valley, Gina Bianchini, Mark Casey, Margit Wennmachers, Robin Chan, Hans Tung, Paul Kranhold, and Om Malik offered insight and advice along the way. I'm also indebted to Anne Kornblut, Michael Jordan, and Ethan Watters for their guidance and, as always, their friendship.

At Bloomberg, John Micklethwait, Reto Gregori, Ellen Pollock, Brad Wieners, Jared Sandberg, and Kristin Powers have provided a wonderful professional home that is committed to ambitious journalism in all forms. I'm also lucky to work at Bloomberg with the best technology team in the business. Tom Giles, Jillian Ward, Peter Elstrom, Nate Lanxon, Aki Ito, Emily Biuso, and Alistair Barr are incredible colleagues who gave me air cover while I was occupied by the pursuit of elusive facts and coherent sentences. Eric Newcomer, Ellen Huet, Mark Milian, Jim Aley, and Max Chafkin read the manuscript early and offered crucial advice. Lulu Chen helped me report the remarkable story of Didi Chuxing in Beijing. Emily Chang was an empathetic confidante as I worked through the challenges of telling this story. And my longtime colleague and co-conspirator Ashlee Vance was, as always, an inspiration and sounding board whenever I needed to find my footing.

Once again, I owe a big thank-you to Nick Sanchez, who ably helped with research and reporting. Several crucial episodes in this account wouldn't be here without his efforts. (Of course, the blunders are all mine.) Diana Suryakusuma assisted with photographs, answering the call even when she was working on the other side of the world.

My family was remarkably patient and helpful during the reporting and writing of this book. Carol Glick, Robert Stone, Luanne Stone, Bernice Yaspan, Brian Stone, Eric Stone, and Becca Stone are now skilled at managing a self-absorbed author. Harper Fox, Maté Schissler, Andrew Iorgulescu, Essence Kelley, and David Lewis were warm and enthusiastic throughout, despite the fact that we root for rival baseball teams.

Even though they would occasionally sigh with exasperation over their distracted daddy, my daughters, Isabella and Calista Stone, offered plenty of motivation and good cheer during this effort.

And finally, writing this book wouldn't have been conceivable without the love and boundless support of Tiffany Fox.

Notes

Unless otherwise noted, quoted material is taken from personal interviews conducted by the author.

Introduction

1. "Extreme Inaugural Experiences," *Good Morning America,* aired January 20, 2009.
2. "Real Time Net Worth," *Forbes,* May 24, 2016, http://www.forbes.com/profile/brian-chesky/; http://www.forbes.com/profile/joe-gebbia/.

Part I
Chapter 1: The Trough of Sorrow

1. "The First Guest Ever on Airbnb Tells His Story," YouTube video, September 20, 2012, https://youtu.be/jpxInV9es6M.
2. Nathaniel Mott, "Watch Our *PandoMonthly* Interview with Airbnb's Brian Chesky," *Pando,* January 11, 2013, https://pando.com/2013/01/11/watch-our-pandomonthly-interview-with-airbnbs-brian-chesky/.
3. Ibid.
4. Episode 109, *American Inventor,* ABC, aired May 4, 2006.
5. Brian Chesky, "View Work by Brian Chesky at Coroflot.com," Coroflot, July 16, 2006, http://www.coroflot.com/brianchesky/view-work.
6. Squirrelbait, "AirBed & Breakfast for Connecting '07," Core77, October 10, 2007, http://www.core77.com/posts/7715/airbed-breakfast-for-connecting-07-7715.
7. Mott, "Watch Our *PandoMonthly* Interview."
8. "Greg McAdoo, Partner at Sequoia Capital, at Startup School '08," YouTube, January 29, 2009, https://www.youtube.com/watch?v=fZ5F2KhMLiE.
9. Brian Chesky, "7 Rejections," *Pulse,* July 13, 2015, https://www.linkedin.com/pulse/7-rejections-brian-chesky.

10. Erick Schonfeld, "AirBed and Breakfast Takes Pad Crashing to a Whole New Level," *TechCrunch,* August 11, 2008, http://techcrunch.com/2008/08/11/airbed-and-breakfast-takes-pad-crashing-to-a-whole-new-level/.
11. Fred Wilson, "Airbnb," *AVC,* March 16, 2011, http://avc.com/2011/03/airbnb/.
12. Paige Craig, "Airbnb, My $1 Billion Lesson," *Arena Ventures,* July 22, 2015. In this blog post, Craig suggests it was Airbnb's entry into Y Combinator that scuttled his deal, but Airbnb got into the program in December, so the timing doesn't quite match up. https://arenavc.com/2015/07/airbnb-my-1-billion-lesson/.
13. Matthew Bandyk, "Republican and Democratic Conventions Still Have Room," *U.S. News and World Report,* August 20, 2008.
14. Lori Rackl, "Airbed & Breakfast, Anyone? New Web Site an Alternative to Pricey, Scarce Hotel Rooms," *Chicago Sun-Times,* August 27, 2008.
15. "Obama O's," Drunkily's Channel, YouTube video, January 12, 2012, https://youtu.be/OQTWimfGfV8.
16. Mott, "Watch Our *PandoMonthly* Interview."

Chapter 2: Jam Sessions

1. "Uber Happy Hour," Vimeo, February 2, 2011, https://vimeo.com/19508742.
2. M. G. Siegler, "StumbleUpon Beats Skype in Escaping eBay's Clutches," *TechCrunch,* April 13, 2009, http://techcrunch.com/2009/04/13/ebay-unacquires-stumbleupon/.
3. "Travis Kalanick, Uber and Loic Le Meur, Co-Founder, LeWeb," YouTube video, December 13, 2013, https://youtu.be/vnkvNQ2V6Og.
4. Siegler, "StumbleUpon Beats Skype."
5. Erin Biba, "Inside the GPS Revolution: 10 Applications That Make the Most of Location," Wired.com, January 19, 2009, http://www.wired.com/2009/01/lp-10coolapps/.
6. "Fireside Chat with Travis Kalanick and Marc Benioff," September 17, 2015, https://www.youtube.com/watch?v=Zt8L8WSSr1g.
7. David Cohen, "The Pony's Lucky Horseshoe," *Hi, I'm David G. Cohen,* July 14, 2014, http://davidgcohen.com/2014/07/14/the-ponys-lucky-horseshoe/.
8. Leena Rao, "UberCab Takes the Hassle Out of Booking a Car Service," *TechCrunch,* July 5, 2010, http://techcrunch.com/2010/07/05/ubercab-takes-the-hassle-out-of-booking-a-car-service/.

Chapter 3: The Nonstarters

1. Jason Kincaid, "Taxi Magic: Hail a Cab from Your iPhone at the Push of a Button," *TechCrunch,* December 16, 2008.

2. Carolyn Said, "DeSoto, S.F.'s Oldest Taxi Firm, Rebrands Itself as Flywheel," *SFGate,* February 19, 2015, http://www.sfgate.com/business/article/DeSoto-S-F-s-oldest-taxi-firm-rebrands-6087480.php.
3. "Why Couchsurfing Founder Casey Fenton Is Unfazed by Competitors like Airbnb," *Mixergy,* March 30, 2015, https://mixergy.com/interviews/casey-fenton-couchsurfing/.
4. Ryan Lawler, "Lyft-Off: Zimride's Long Road to Overnight Success," *TechCrunch,* August 29, 2014, http://techcrunch.com/2014/08/29/6000-words-about-a-pink-mustache/.
5. "Cross-Country Carpool," *ABC News,* July 29, 2008, http://abcnews.go.com/video/embed?id=5456748.

Chapter 4: The Growth Hacker

1. Nathaniel Mott, "Watch Our *PandoMonthly* Interview with Airbnb's Brian Chesky," *Pando,* January 11, 2013, https://pando.com/2013/01/11/watch-our-pandomonthly-interview-with-airbnbs-brian-chesky/.
2. "Reid Hoffman and Brian Chesky (11/2/11)," YouTube video, November 15, 2011, https://youtu.be/dPp9zc6SIHY.
3. Ibid.
4. Ibid.
5. "Data-Miners.net—Nathan Blecharczyk," Spamhaus, http://archive.org/web/20030512215519/http://www.spamhaus.org/rokso/spammers.lasso?-database=spammers.db&-layout=detail&-response=roksodetail.lasso&recno=2259&-clientusername=guest&-clientpassword=guest&-search.
6. Aaron Greenspan, "The Harvard People I Know Who Are Breaking the Law (Again)," October 26, 2011, https://thinkcomp.quora.com/The-Harvard-People-I-Know-Who-Are-Breaking-The-Law-Again.
7. "ComScore Media Metrix Ranks Top 50 U.S. Web Properties for October 2009," ComScore, November 19, 2009.
8. Dave Gooden, "How Airbnb Became a Billion-Dollar Company," May 31, 2011, http://davegooden.com/2011/05/how-airbnb-became-a-billion-dollar-company/.
9. Ryan Tate, "Did Airbnb Scam Its Way to $1 Billion?," *Gawker,* May 31, 2011, http://gawker.com/5807189/did-airbnb-scam-its-way-to-1-billion.
10. Andrew Chen, "Growth Hacker Is the New VP Marketing," http://andrewchen.co/how-to-be-a-growth-hacker-an-airbnbcraigslist-case-study/.
11. "Airbnb Announces New Product Advancements and $7.2M in Series A Funding to Accelerate Global Growth," *Marketwired,* November 11, 2010, http://www

.marketwired.com/press-release/Airbnb-Announces-New-Product-Advancements-72M-Series-A-Funding-Accelerate-Global-Growth-1351692.htm.

12. "Reid Hoffman and Brian Chesky," YouTube video.
13. Brad Stone, "The New Andreessen," Bloomberg.com, November 3, 2010, http://www.bloomberg.com/news/articles/2010-11-03/the-new-new-andreessen.

Chapter 5: Blood, Sweat, and Ramen

1. "Disrupt Backstage: Travis Kalanick," YouTube video, June 22, 2011, https://youtu.be/0-uiO-P9yEg.
2. Ilene Lelchuk, "Probe Clears 2 S.F. Elections Officials; Case Against 3rd Remains Unclear," *SFGate,* December 12, 2001, http://www.sfgate.com/politics/article/Probe-clears-2-S-F-elections-officials-Case-2841381.php.
3. Andy Kessler, "Travis Kalanick: The Transportation Trustbuster," *Wall Street Journal,* January 25, 2013, http://www.wsj.com/articles/SB10001424127887324235104578244231122376480.
4. "Disrupt Backstage: Travis Kalanick," YouTube video.
5. "Travis Kalanick Startup Lessons from the Jam Pad—Tech Cocktail Startup Mixology," YouTube video, May 5, 2011, https://youtu.be/VMvdvP02f-Y.
6. Max Chafkin, "What Makes Uber Run," *Fast Company,* September 8, 2015, http://www.fastcompany.com/3050250/what-makes-uber-run.
7. Ibid.
8. "Travis Kalanick Startup Lessons from the Jam Pad."
9. "Travis Kalanick of Uber," *This Week in Startups,* YouTube video, August 16, 2011, https://youtu.be/550X5OZVk7Y.
10. "Power Tools," *Time,* April 24, 2014, http://time.com/72206/time-100-objects-that-inspire-influencers/.
11. Karen Kaplan, "Ovitz Team Invests in Multimedia Search Engine," *Los Angeles Times,* June 10, 1999, http://articles.latimes.com/1999/jun/10/business/fi-46036.
12. Bruce Orwall, "Ovitz, Yucaipa Buy Majority Stake in Entertainment Search Engine," *Wall Street Journal,* June 10, 1999, http://www.wsj.com/articles/SB928970934179363266.
13. Ibid.
14. Marc Graser and Justin Oppelaar, "Scour Power Turns H'wood Dour," *Variety,* June 24, 2000.
15. "Travis Kalanick of Uber," *This Week in Startups.*

16. Karen Kaplan and P. J. Huffstutter, "Multimedia Firm Scour Lays Off 52 of Its 70 Workers," *Los Angeles Times,* September 2, 2000, http://articles.latimes.com/2000/sep/02/business/fi-14350.
17. Clare Saliba, "Scour Assets Sell for $9M," *E-Commerce Times,* December 13, 2000, http://www.ecommercetimes.com/story/6043.html.
18. "FailCon 2011—Uber Case Study," YouTube video, November 3, 2011, https://youtu.be/2QrX5jsiico.
19. Ibid.
20. Ibid.
21. Travis Kalanick, interview by Ashlee Vance, September 30, 2011.
22. "FailCon 2011—Uber Case Study," YouTube video.
23. Ibid.
24. Michael Arrington, "Payday for Red Swoosh: $15 Million from Akamai," *TechCrunch,* April 12, 2007, http://techcrunch.com/2007/04/12/payday-for-red-swoosh-15-million-from-akamai/.
25. Author's interview with Travis Kalanick and http://fortune.com/2013/09/19/travis-kalanick-founder-of-uber-is-silicon-valleys-rebel-hero/.
26. "Travis Kalanick Startup Lessons from the Jam Pad," YouTube video.
27. Ibid.
28. "Travis Kalanick of Uber," *This Week in Startups.*
29. Ryan Graves, "1 + 1 = 3," Uber.com, December 22, 2010, https://newsroom.uber.com/1-1-3/.

Part II

Chapter 6: The Wartime CEO

1. Ben Horowitz, "Peacetime CEO/Wartime CEO," *Ben's Blog,* Andreessen Horowitz, April 14, 2011, http://www.bhorowitz.com/peacetime_ceo_wartime_ceo.
2. Aileen Lee, "Welcome to the Unicorn Club: Learning from Billion-Dollar Startups," *TechCrunch,* November 2, 2013, https://techcrunch.com/2013/11/02/welcome-to-the-unicorn-club/.
3. Glenn Peoples, "Spotify Raises $100 Million, but Remains Stuck at $1 Billion Valuation," *Billboard,* June 17, 2011, http://www.billboard.com/biz/articles/news/1177428/spotify-raises-100-million-but-remains-stuck-at-1-billion-valuation; Geoffrey Fowler, "Airbnb Is Latest Start-Up to Secure $1 Billion Valuation," *Wall Street Journal,* July 26, 2011, http://www.wsj.com/articles/SB10001424053111904772304576468183971793712.

4. Eric Mack, "Plane-in-a-Tree Is the Perfect Getaway for Airbnb," CNET.com, August 1, 2011, http://www.cnet.com/news/plane-in-a-tree-is-the-perfect-getaway-for-airbnb/.
5. "How Airbnb and Uber Disrupt Offline Business," *TechCrunch,* December 28, 2011, http://techcrunch.com/video/how-airbnb-and-uber-disrupt-offline-business/517158889/.
6. Sarah Lacy, "Airbnb Has Arrived: Raising Mega-Round at a $1 Billion+ Valuation," *TechCrunch,* May 30, 2011, http://techcrunch.com/2011/05/30/airbnb-has-arrived-raising-mega-round-at-a-1-billion-valuation/.
7. Steve O'Hear, "9flats, the European Airbnb, Secures 'Major Investment' from Silicon Valley's Redpoint," *TechCrunch,* May 17, 2011, http://techcrunch.com/2011/05/17/9flats-the-european-airbnb-secures-major-investment-from-silicon-valleys-redpoint-2/.
8. "Attack of the Clones," *Economist,* August 6, 2011, http://www.economist.com/node/21525394.
9. Mike Butcher, "In Confidential Email Samwer Describes Online Furniture Strategy as a 'Blitzkrieg,'" *TechCrunch,* December 22, 2011, http://techcrunch.com/2011/12/22/in-confidential-email-samwer-describes-online-furniture-strategy-as-a-blitzkrieg/.
10. Caroline Winter, "How Three Germans Are Cloning the Web," *Bloomberg,* February 29, 2012, http://www.bloomberg.com/news/articles/2012-02-29/how-three-germans-are-cloning-the-web.
11. Robin Wauters, "Investors Pump $90 Million into Airbnb Clone Wimdu," *TechCrunch,* June 14, 2011, http://techcrunch.com/2011/06/14/investors-pump-90-million-into-airbnb-clone-wimdu/.
12. EJ, "Violated: A Traveler's Lost Faith, a Difficult Lesson Learned," *Around the World and Back Again,* June 29, 2011, http://ejroundtheworld.blogspot.com/2011/06/violated-travelers-lost-faith-difficult.html.
13. Ibid.
14. Foxit, "Violated: A Traveler's Lost Faith, a Difficult Lesson Learned," Hacker News, https://news.ycombinator.com/item?id=2811080.
15. Michael Arrington, "The Moment of Truth for Airbnb As User's Home Is Utterly Trashed," *TechCrunch,* July 27, 2011, http://techcrunch.com/2011/07/27/the-moment-of-truth-for-airbnb-as-users-home-is-utterly-trashed/.
16. EJ, "Airbnb Nightmare: No End in Sight," *Around the World and Back Again,* July 28, 2011, http://ejroundtheworld.blogspot.com/2011/07/airbnb-nightmare-no-end-in-sight.html.
17. Ibid.

18. Drew Olanoff, "Airbnb Ups Its Host Guarantee to a Million Dollars," *Next Web,* May 22, 2012, http://thenextcom/insider/2012/05/22/airbnb-partners-with-lloyds-of-london-for-the-new-million-dollar-host-guarantee/.
19. Brian Chesky, "Our Commitment to Trust & Safety," Airbnb, August 1, 2011, http://blog.airbnb.com/our-commitment-to-trust-and-safety/.
20. James Temple, "Airbnb Victim Describes Crime and Aftermath," *SFGate,* July 30, 2011, http://www.sfgate.com/business/article/Airbnb-victim-describes-crime-and-aftermath-2352693.php.
21. Claire Cain Miller, "In Silicon Valley, the Night Is Still Young," *New York Times,* August 20, 2011, http://www.nytimes.com/2011/08/21/technology/silicon-valley-booms-but-worries-about-a-new-bust.html.
22. Jim Wilson, "Good Times in Silicon Valley, for Now," *New York Times,* August 13, 2011, http://www.nytimes.com/slideshow/2011/08/13/technology/20110821-VALLEY-5.html; Geoffrey Fowler, "The Perk Bubble Is Growing as Tech Booms Again," *Wall Street Journal,* July 6, 2011, http://www.wsj.com/articles/SB10001424052702303763404576419803997423690.
23. Robin Wauters, "Airbnb Buys German Clone Accoleo, Opens First European Office in Hamburg," *TechCrunch,* June 1, 2011, http://techcrunch.com/2011/06/01/airbnb-buys-german-clone-accoleo-opens-first-european-office-in-hamburg/.
24. Colleen Taylor, "Airbnb Hits Hockey Stick Growth: 10 Million Nights Booked, 200K Active Properties," *TechCrunch,* June 19, 2012, http://techcrunch.com/2012/06/19/airbnb-10-million-bookings-global/.

Chapter 7: The Playbook

1 Erick Schonfeld, "I Just Rode in an Uber Car in New York City, and You Can Too," *TechCrunch,* April 6, 2011, http://techcrunch.com/2011/04/06/i-just-rode-in-an-uber-car-in-new-york-city-and-you-can-too/.
2. Andrew J. Hawkins, "Uber Doubles Number of Drivers—Just as De Blasio Feared," *Crain's New York Business,* October 6, 2015, http://www.crainsnewyork.com/article/20151006/BLOGS04/151009912/uber-doubles-number-of-drivers-just-as-de-blasio-feared.
3. Nitasha Tiku, "Exclusive: Shake Up and Resignations at Uber's New York Office, CEO Travis Kalanick Explains," *Observer,* September 20, 2011, http://observer.com/2011/09/exclusive-shake-up-and-resignations-at-ubers-new-york-office-ceo-travis-kalanick-explains/.
4. Full disclosure: Bloomberg is my employer!

5. "Travis Kalanick of Uber," *This Week in Startups,* YouTube video, August 16, 2011, https://youtu.be/550X5OZVk7Y.
6. "Halloween Surge Pricing: Get an Uber at the Witching Hour," Uber, October 26, 2011, https://newsroom.uber.com/halloween-surge-pricing-get-an-uber-at-the-witching-hour/.
7. Aubrey Sabala, "While I'm Glad I'm Home Safely," Twitter, January 1, 2012, https://twitter.com/aubs/status/153532514122743808.
8. Travis Kalanick, "@kavla Price Is Right There Before You Request," Twitter, January 2, 2012, https://twitter.com/travisk/status/154069401488982017.
9. Travis Kalanick, "@dandarcy the Sticker Shock Is Rough," Twitter, January 1, 2012, https://twitter.com/travisk/status/153562288023023617.
10. Nick Bilton, "Disruptions: Taxi Supply and Demand, Priced by the Mile," *Bits Blog, New York Times,* January 8, 2012, http://bits.blogs.nytimes.com/2012/01/08/disruptions-taxi-supply-and-demand-priced-by-the-mile/?_r=0.
11. Bill Gurley, "A Deeper Look at Uber's Dynamic Pricing Model," *Above the Crowd,* March 11, 2014, http://abovethecrowd.com/2014/03/11/a-deeper-look-at-ubers-dynamic-pricing-model/.
12. Kara Swisher, "Man and Uber Man," *Vanity Fair,* December 2014, http://www.vanityfair.com/news/2014/12/uber-travis-kalanick-controversy.
13. Alex Konrad, "How Super Angel Chris Sacca Made Billions, Burned Bridges and Crafted the Best Seed Portfolio Ever," *Forbes,* March 25, 2015, http://www.forbes.com/sites/alexkonrad/2015/03/25/how-venture-cowboy-chris-sacca-made-billions/#5d29290bfa8c.

Chapter 8: Travis's Law

1. Travis Kalanick, "Uber CEO's Letter to DC City Council," Uber, July 10, 2012, https://newsroom.uber.com/us-dc/travis-kalanick-letter-to-dc-city-council/.
2. Benjamin R. Freed, "Uber Is Hacking into Washington's Taxi Industry, Linton Says," *DCist,* January 11, 2012, http://dcist.com/2012/01/uber_is_hacking_into_washingtons_ta.php.
3. "D.C. Regulations on Limousine Operators," Scribd, https://www.scribd.com/doc/77931261/D-C-Regulations-on-Limousine-Operators.
4. Mike DeBonis, "Uber Car Impounded, Driver Ticketed in City Sting," *Washington Post,* January 13, 2012, https://www.washingtonpost.com/blogs/mike-debonis/post/uber-car-impounded-driver-ticketed-in-city-sting/2012/01/13/gIQA4Py3vP_blog.html.
5. Ryan Graves, "An Uber Surprise in DC," Uber, January 13, 2012, https://newsroom.uber.com/us-dc/an-uber-surprise-in-dc/.

6. Benjamin R. Freed, "After Stinging Uber, Linton Says He Just Had to Regulate," *DCist,* January 16, 2012, http://dcist.com/2012/01/after_stinging_uber_linton_says_he.php.
7. Leena Rao, "Mobile Taxi Network Hailo Raises $17M From Accel and Atomico to Take On Uber in the U.S.," *TechCrunch,* March 29, 2012, http://techcrunch.com/2012/03/29/mobile-taxi-network-hailo-raises-17m-from-accel-and-atomico-to-take-on-uber-in-the-u-s/.
8. Ibid.
9. Laura June, "Uber Launches Lower-Priced Taxi Service in Chicago," *Verge,* April 18, 2012, http://www.theverge.com/2012/4/18/2957508/uber-taxi-service-chicago.
10. Daniel Cooper, "Hailo's HQ Trashed by Uber-Hating London Black Cab Drivers," *Engadget,* May 23, 2014, https://www.engadget.com/2014/05/23/hailo-london-hq-vandalized/.
11. "SF, You Now Have the Freedom to Choose," Uber, July 3, 2012, http://blog.uber.com/2012/07/03/sf-vehicle-choice/.
12. Brian X. Chen, "Uber, an App That Summons a Car, Plans a Cheaper Service Using Hybrids," *New York Times*, July 1, 2012, http://www.nytimes.com/2012/07/02/technology/uber-a-car-service-smartphone-app-plans-cheaper-service.html.
13. Mike DeBonis, "Uber CEO Travis Kalanick," *Washington Post,* July 27, 2012, https://www.washingtonpost.com/blogs/mike-debonis/post/uber-ceo-travis-kalanick-talks-big-growth-and-regulatory-roadblocks-in-dc/2012/07/27/gJQAAmS4DX_blog.html.
14. Del Quentin Wilber and Mike DeBonis, "Ted G. Loza, Former D.C. Council Aide, Pleads Guilty in Corruption Case," *Washington Post,* February 18, 2011, http://www.washingtonpost.com/wp-dyn/content/article/2011/02/18/AR2011021806843.html.
15. Travis Kalanick, "@mikedebonis We Felt That We Got Strung Out," Twitter, July 10, 2012, https://twitter.com/travisk/status/222633686770786305; Travis Kalanick, "@mikedebonis the Bottom Line Is That @marycheh," Twitter, July 10, 2012, https://twitter.com/travisk/status/222635403910447104.
16. Travis Kalanick, "Strike Down the Minimum Fare Language in the DC Uber Amendment," Uber, July 9, 2012, https://newsroom.uber.com/us-dc/strike-down-the-minimum-fare/.
17. Christine Lagorio-Chafkin, "Resistance Is Futile," *Inc.,* July 2013, http://www.inc.com/magazine/201307/christine-lagorio/uber-the-car-service-explosive-growth.html.

18. Mike DeBonis, "Uber Triumphant," *Washington Post,* December 2012, https://www.washingtonpost.com/blogs/mike-debonis/wp/2012/12/03/uber-triumphant/.
19. "Patent US6356838—System and Method for Determining an Efficient Transportation Route," March 12, 2002, http://www.google.com/patents/US6356838.
20. There were other ridesharing companies that preceded Sidecar. Starting in 2010, one San Francisco service, called Homobile, offered transvestite performers and members of the gay community rides and solicited donations as payment. Sunil Paul says he tried the service on a trip to the airport in 2011.
21. "Travis Kalanick of Uber," *This Week in Startups,* YouTube video, August 16, 2011, https://youtu.be/550X5OZVk7Y.
22. Tomio Geron, "Ride-Sharing Startups Get California Cease and Desist Letters," *Forbes,* October 8, 2012, http://www.forbes.com/sites/tomiogeron/2012/10/08/ride-sharing-startups-get-california-cease-and-desist-letters/#767d66027e81.
23. Jeff McDonald and Ricky Young, "State Investigator Lays Out Developing Criminal Case Against Former PUC President," *Los Angeles Times,* December 29, 2015, http://www.latimes.com/business/la-fi-watchdog-peevey-20151230-story.html.
24. Sfcda.com/CPUC, January 11, 2013, http://sfcda.com/CPUC/Lyft_CPUC_SED_IntAGR.pdf.
25. Brian X. Chen, "Uber to Roll Out Ride Sharing in California," *Bits Blog, New York Times,* January 31, 2013, http://bits.blogs.nytimes.com/2013/01/31/uber-rideshare/.
26. Travis Kalanick, "@johnzimmer You've Got a Lot of Catching Up," Twitter, March 19, 2013, https://twitter.com/travisk/status/314079323478962176.
27. David Pierson, "Uber Fined $7.6 Million by California Utilities Commission," *Los Angeles Times,* January 14, 2016, http://www.latimes.com/business/la-fi-tn-uber-puc-20160114-story.html.
28. "Order Instituting Rulemaking on Regulations Relating to Passenger Carriers, Ridesharing, and New Online-Enabled Transportation Services," Cpuc.ca.gov, September 19, 2013, http://docs.cpuc.ca.gov/PublishedDocs/Published/G000/M077/K112/77112285.PDF.
29. Liz Gannes, "Despite Controversy in Austin and Philly, Ride-Sharing Service SideCar Expands to Boston, Brooklyn and Chicago," *AllThingsD,* March 15, 2013, http://allthingsd.com/20130315/despite-controversy-in-austin-and-philly-ride-sharing-service-sidecar-expands-to-boston-brooklyn-and-chicago/.

Chapter 9: Too Big to Regulate

1. Sarah Kessler, "How Snow White Helped Airbnb's Mobile Mission," *Fast Company,* November 8, 2012, http://www.fastcompany.com/3002813/how-snow-white-helped-airbnbs-mobile-mission.

2. Kristen Bellstrom, "Exclusive: Meet Airbnb's Highest-Ranking Female Exec Ever," *Fortune,* July 13, 2015, http://fortune.com/2015/07/13/airbnb-belinda-johnson-promotion/.
3. Nicole Neroulias, "Fan 'Gridderati' Get Super Soiree—Sexy Treat at Top-of-Line Bash," *New York Post,* February 4, 2007.
4. Justin Rocket Silverman, "He's King of the City That Never Sleeps," *AM New York,* June 24, 2004.
5. Ben Chapman, "Website AirBnB.com Lets Users Sublet Couches, Roofs and Other Odd Spaces," *New York Daily News,* July 21, 2009, http://www.nydailynews.com/life-style/real-estate/website-airbnb-lets-users-sublet-couches-roofs-odd-spaces-article-1.429969.
6. Ibid.
7. Joe Gebbia, "I'll Be in NYC Tomorrow," Twitter, July 20, 2010, https://twitter.com/jgebbia/status/19046704645.
8. "Statements of Mayor Michael R. Bloomberg and Governor David A. Paterson on Governor Paterson's Signing Into Law Housing Preservation Legislation That Enables Enforcement Against Illegal Hotels," City of New York, July 23, 2010, http://www1.nyc.gov/office-of-the-mayor/news/324-10/statements-mayor-michael-bloomberg-governor-david-a-paterson-governor-paterson-s.
9. Andrew J. Hawkins, "City Sues Departed Actor for Running Illegal Hotels," *Crain's New York Business,* October 23, 2012, http://www.crainsnewyork.com/article/20121023/BLOGS04/310239983/city-sues-departed-actor-for-running-illegal-hotels.
10. Drew Grant, "Infamous Airbnb Hotelier Toshi to Pay $1 Million to NYC," *Observer,* November 20, 2013, http://observer.com/2013/11/infamous-airbnb-hotelier-toshi-to-pay-1-million-to-nyc/.
11. Adam Pincus, "Illegal Hotel Fines Could Skyrocket," *Real Deal,* September 12, 2012, http://therealdeal.com/2012/09/12/city-council-to-dramatically-increase-illegal-hotel-fines/.
12. Ron Lieber, "A Warning for Hosts of Airbnb Travelers," *New York Times,* November 30, 2012, http://www.nytimes.com/2012/12/01/your-money/a-warning-for-airbnb-hosts-who-may-be-breaking-the-law.html?_r=1.
13. *NYC v. Abe Carrey Appeal Nos. 1300602 & 1300736,* CityLaw.org, September 26, 2013, http://archive.citylaw.org/ecb/Long%20Form%20Orders/2013/1300602—1300736.pdf.
14. "Huge Victory in New York for Nigel Warren and Our Host Community," Airbnb, September 27, 2013, https://www.airbnbaction.com/huge-victory-new-york-nigel-warren-host-community/.

15. Brian Chesky, "Who We Are, What We Stand For," Airbnb, October 3, 2013, http://blog.airbnb.com/who-we-are/.
16. http://valleywag.gawker.com/airbnb-hides-warning-that-users-are-breaking-the-law-in-1561938121.
17. Matt Chaban, "Attorney General Eric Schneiderman Hits AirBnB with Subpoena for User Data," *New York Daily News,* October 7, 2013, http://www.nydailynews.com/news/national/state-airbnb-article-1.1477934.
18. "Airbnb Memorandum in Support of Petition to Quash Subpoena," Electronic Frontier Foundation, https://www.eff.org/document/airbnb-v-schneiderman-memo-law.
19. "Airbnb Introduces Instant Bookings for Hosts," ProBnB, October 12, 2013, http://www.probnb.com/airbnb-introduces-instant-bookings-for-hosts.
20. Daniel P. Tucker, "Airbnb Won't Comply with Subpoena from New York Attorney General," WNYC, October 7, 2013, http://www.wnyc.org/story/airbnb-wont-comply-subpoena-new-york-attorney-general/.
21. "Airbnb's Economic Impact on the NYC Community," Airbnb, http://blog.airbnb.com/airbnbs-economic-impact-nyc-community/.
22. "Ruling in Airbnb's Case in New York," *New York Times,* May 13, 2014, http://www.nytimes.com/interactive/2014/05/13/technology/ruling-airbnb-new-york.html.
23. "New York Update," Airbnb, August 22, 2014, https://www.airbnbaction.com/new-york-community-update/.
24. http://www.ag.ny.gov/press-release/ag-schneiderman-releases-report-documenting-widespread-illegality-across-airbnbs-nyc.
25. Jessica Wohl, "Airbnb CMO Knocks Uber's Growth Tactics," *Advertising Age,* October 16, 2015, http://adage.com/article/special-report-ana-annual-meeting-2015/airbnb-cmo-knocks-uber-growth-tactics/300948/.

Part III

Chapter 10: God View

1. Jeanie Riess, "Why New Orleans Doesn't Have Uber," *Gambit,* February 4, 2014, http://www.bestofneworleans.com/gambit/why-new-orleans-doesnt-have-uber/Content?oid=2307943.
2. Tim Elfrink, "UberX Will Launch in Miami Today, Defying Miami-Dade's Taxi Laws," *Miami New Times,* June 4, 2014, http://www.miaminewtimes.com/news/uberx-will-launch-in-miami-today-defying-miami-dades-taxi-laws-6533024.

3. "Mayor Gimenez: Uber, Lyft Will Be Legal in Miami-Dade by End of Year," *Miami Herald,* September 28, 2015, http://www.miamiherald.com/news/local/community/miami-dade/article36831345.html.
4. Leena Rao, "Uber Now Offers Its Own Car Leases to UberX Drivers," *Forbes,* July 29, 2015, http://fortune.com/2015/07/29/uber-car-leases/.
5. Travis Kalanick, interview with Mark Milian, November 22, 2013.
6. Eric Newcomer and Olivia Zaleski, "Inside Uber's Auto-Lease Machine, Where Almost Anyone Can Get a Car," Bloomberg.com, May 31, 2016, http://www.bloomberg.com/news/articles/2016-05-31/inside-uber-s-auto-lease-machine-where-almost-anyone-can-get-a-car.
7. Ryan Lawler, "Uber Slashes UberX Fares in 16 Markets to Make It the Cheapest Car Service Available Anywhere," *TechCrunch,* January 9, 2014, http://techcrunch.com/2014/01/09/big-uberx-price-cuts/.
8. Ellen Huet, "How Uber and Lyft Are Trying to Kill Each Other," *Forbes,* May 30, 2014, http://www.forbes.com/sites/ellenhuet/2014/05/30/how-uber-and-lyft-are-trying-to-kill-each-other/#4a7e6b063ba8.
9. Carolyn Tyler, "Mother of Girl Fatally Struck by Uber Driver Speaks Out," *ABC7 News,* December 9, 2014, http://abc7news.com/business/mother-of-girl-fatally-struck-by-uber-driver-speaks-out/429535/.
10. Travis Kalanick, "@connieezywe Can Confirm," Twitter, January 1, 2014, https://twitter.com/travisk/status/418518282824458241.
11. "Statement on New Year's Eve Accident," Uber, January 1, 2014, https://newsroom.uber.com/statement-on-new-years-eve-accident/.
12. Elyce Kirchner, David Paredes, and Scott Pham, "UberX Driver in Fatal Crash Had Record," *NBC Bay Area,* February 12, 2015, http://www.nbcbayarea.com/news/local/UberX-Driver-Involved-in-New-Years-Eve-Manslaughter-Had-A-Record-of-Reckless-Driving-240344931.html.
13. "Fact Sheet 16a: Employment Background Checks in California: A Focus on Accuracy," Privacy Rights Clearinghouse, 2003–2016, https://www.privacyrights.org/employment-background-checks-california-focus-accuracy.
14. Don Jergler, "Uber Announces New Policy to Cover Gap," *Insurance Journal,* March 14, 2014, http://www.insurancejournal.com/news/national/2014/03/14/323329.htm.
15. Harrison Weber, "Uber & Lyft Agree to Insure Drivers in Between Rides in California," *VentureBeat,* August 27, 2014, http://venturebeat.com/2014/08/27/uber-lyft-agree-to-insure-drivers-in-between-rides-in-california/.
16. Bob Egelko, "Uber May Be Liable for Accidents, Even If Drivers Are Contractors," *San Francisco Chronicle,* April 27, 2016, http://www.sfchronicle

.com/bayarea/article/Uber-may-be-liable-for-accidents-even-if-drivers-7377364.php.

17. "Family of 6-Year-Old Girl Killed by Uber Driver Settles Lawsuit," *ABC7 News,* July 14, 2015, http://abc7news.com/business/family-of-6-year-old-girl-killed-by-uber-driver-settles-lawsuit/852108/.
18. "Uber's Marketing Program to Recruit Drivers: Operation SLOG," Uber, August 26, 2014, https://newsroom.uber.com/ubers-marketing-program-to-recruit-drivers-operation-slog/.
19. Laurie Segall, "Uber Rival Accuses Car Service of Dirty Tactics," *CNN Money,* January 24, 2014, http://money.cnn.com/2014/01/24/technology/social/uber-gett/.
20. Mickey Rapkin, "Uber Cab Confessions," *GQ,* February 27, 2014, http://www.gq.com/story/uber-cab-confessions.
21. Ryan Lawler, "Lyft Launches in 24 New Markets, Cuts Fares by Another 10%," *TechCrunch,* April 24, 2014, https://techcrunch.com/2014/04/24/lyft-24-new-cities/.
22. Kara Swisher, "Man and Uber Man," *Vanity Fair,* December 2014, http://www.vanityfair.com/news/2014/12/uber-travis-kalanick-controversy.
23. Sara Ashley O'Brien, "15 Questions with... John Zimmer," CNN, http://money.cnn.com/interactive/technology/15-questions-with-john-zimmer/.
24. Yuliya Chernova, "N.Y. Shutdowns for SideCar, RelayRides Highlight Hurdles for Car- and Ride-Sharing Startups," *Wall Street Journal,* May 15, 2013, http://blogs.wsj.com/venturecapital/2013/05/15/n-y-shutdowns-for-sidecar-relayrides-highlight-hurdles-for-car-and-ride-sharing-startups/.
25. "Lyft Will Launch in Brooklyn & Queens," *Lyft Blog,* July 8, 2014, https://blog.lyft.com/posts/2014/7/8/lyft-launches-in-new-yorks-outer-boroughs.
26. Brady Dale, "Lyft Launch Party with Q-Tip, Without Actually Launching," *Technical.ly Brooklyn,* July 14, 2014, http://technical.ly/brooklyn/2014/07/14/lyft-brooklyn-launches/.
27. "Lyft Launches in NYC," *Lyft Blog,* July 25, 2014, https://blog.lyft.com/posts/2014/7/25/lyft-launches-in-nyc.
28. Casey Newton, "This Is Uber's Playbook for Sabotaging Lyft," *Verge,* August 26, 2014, http://www.theverge.com/2014/8/26/6067663/this-is-ubers-playbook-for-sabotaging-lyft.
29. In September 2012, I washed cars for the Cherry service in San Francisco and was mentored and reviewed by an older washer, Kenny Chen. "Brad needs to look out for traffic," he wrote; Brad Stone, "My Life as a TaskRabbit," Bloomberg.com, September 13, 2012, http://www.bloomberg.com/news/articles/2012-09-13/my-life-as-a-taskrabbit.

30. Dan Levine, "Exclusive: Lyft Board Members Discussed Replacing CEO, Court Documents Reveal," Reuters, November 7, 2014, http://www.reuters.com/article/us-lyft-ceo-lawsuit-exclusive-idUSKBN0IR2HA20141108.
31. Douglas Macmillan, "Lyft Alleges Former Executive Took Secret Documents with Him to Uber," *Wall Street Journal,* November 5, 2014, http://blogs.wsj.com/digits/2014/11/05/lyft-alleges-former-executive-took-secret-documents-with-him-to-uber/.
32. Travis VanderZanden, "All the Facts Will Come Out," Twitter, November 6, 2014, https://twitter.com/travisv/status/530398592968585217.
33. Joseph Menn and Dan Levine, "Exclusive—U.S. Justice Dept. Probes Data Breach at Uber: Sources," Reuters, December 18, 2015, http://www.reuters.com/article/uber-tech-lyft-probe-exclusive-idUSKBN0U12FH20151219.
34. Dan Levine, "Uber, Lyft Settle Litigation Involving Top Executives," Reuters, June 28, 2016, http://www.reuters.com/article/us-uber-lyft-idUSKCN0ZE0FP.
35. Kristen V. Brown, "Uber Shifts into Mid-Market Headquarters," *San Francisco Chronicle,* June 2, 2014, http://www.sfgate.com/technology/article/Uber-shifts-into-Mid-Market-headquarters-5521166.php.
36. Mike Isaac, "Uber Picks David Plouffe to Wage Regulatory Fight," *New York Times,* August 19, 2014, http://www.nytimes.com/2014/08/20/technology/uber-picks-a-political-insider-to-wage-its-regulatory-battles.html.
37. Kim Lyons, "In Clash of Cultures, PUC Grapples with Brave New Tech World," *Pittsburgh Post-Gazette,* August 24, 2014, http://www.post-gazette.com/business/2014/08/24/In-clash-of-cultures-PUC-grapples-with-brave-new-tech-world/stories/201408240002.
38. Sarah Lacy, "The Horrific Trickle-Down of Asshole Culture: Why I've Just Deleted Uber from My Phone," *Pando,* October 22, 2014, https://pando.com/2014/10/22/the-horrific-trickle-down-of-asshole-culture-at-a-company-like-uber/.
39. Charlie Warzel, "Sexist French Uber Promotion Pairs Riders with 'Hot Chick' Drivers," *BuzzFeed,* October 21, 2014, https://www.buzzfeed.com/charliewarzel/french-uber-bird-hunting-promotion-pairs-lyon-riders-with-a?utm_term=.smxR9a9Q8#.miaNnpnDJ.
40. Lacy, "The Horrific Trickle-Down of Asshole Culture."
41. Ben Smith, "Uber Executive Suggests Digging Up Dirt on Journalists," *BuzzFeed,* November 17, 2014, https://www.buzzfeed.com/bensmith/uber-executive-suggests-digging-up-dirt-on-journalists?utm_term=.dqX1DyDkz#.epX2XQXbO.

42. Nicole Campbell, "What Was Said at the Uber Dinner," *Huffington Post,* November 21, 2014, http://www.huffingtonpost.com/nicole-campbell/what-was-said-at-the-uber_b_6198250.html.

Chapter 11: Escape Velocity

1. Brian Chesky, speech at iHub, Nairobi, July 26, 2015, https://www.youtube.com/watch?v=UFhwh3Ex6Zg.
2. Harrison Weber, "Top Designers React to Airbnb's Controversial New Logo," *VentureBeat,* July 18, 2014, http://venturebeat.com/2014/07/18/top-designers-react-to-airbnbs-controversial-new-logo/.
3. "State of the Airbnb Union: A Keynote with Brian Chesky," YouTube video, November 24, 2014, https://youtu.be/EKX5W8r0Pgc?list=PLe_YVMnSlo XYMnclJtn2-anpH7PDUFip_.
4. Alex Konrad, "Airbnb Cofounders to Become First Sharing Economy Billionaires As Company Nears $10 Billion Valuation," *Forbes,* March 20, 2014, http://www.forbes.com/sites/alexkonrad/2014/03/20/airbnb-cofounders-are-billionaires/#2a6b41b641ab.
5. James Lo Chi-hao, "Backpacker Dies from Carbon Monoxide Poisoning," *China Post,* December 31, 2013, http://www.chinapost.com.tw/taiwan/national/national-news/2013/12/31/397194/Backpacker-dies.htm.
6. Hope Well, "@bchesky Our Daughter Elizabeth Passed Away," Twitter, January 21, 2014, https://twitter.com/hopewell828/status/425777540624424960.
7. William B. Smith, "Taming the Digital Wild West," Abramson Smith Waldsmith, http://www.aswllp.com/content/images/Taming-The-Digital-Wild-West.pdf.
8. Ryan Lawler, "To Ensure Guest Safety, Airbnb Is Giving Away Safety Cards, First Aid Kits, and Smoke & CO Detectors," *TechCrunch,* February 21, 2014, https://techcrunch.com/2014/02/21/airbnb-safety-giveaway/.
9. "A Huge Step Forward for Home Sharing in Portland," Airbnb Action, July 30, 2014, https://www.airbnbaction.com/home-sharing-in-portland/.
10. Elliot Njus, "Airbnb, Acting as Portland's Lodging Tax Collector, Won't Hand Over Users' Names or Addresses," *Oregonian,* July 21, 2014, http://www.oregonlive.com/front-porch/index.ssf/2014/07/airbnb_acting_as_portlands_lod.html.
11. Brian Chesky, "Shared City," *Medium,* March 26, 2014, https://medium.com/@bchesky/shared-city-db9746750a3a.
12. John Cote, "Airbnb, Other Sites Owe City Hotel Tax, S.F. Says," *SFGate,* April 4, 2012, http://www.sfgate.com/bayarea/article/Airbnb-other-sites-owe-city-hotel-tax-S-F-says-3457290.php.

13. Chesky, "Shared City."
14. "San Francisco, Taxes and the Airbnb Community," Airbnb Action, March 31, 2014, https://www.airbnbaction.com/san-francisco-taxes-airbnb-community/.
15. Philip Matier and Andrew Ross, "Airbnb Pays Tax Bill of 'Tens of Millions' to S.F," *SFGate,* February 18, 2015, http://www.sfgate.com/bayarea/matier-ross/article/M-R-Airbnb-pays-tens-of-millions-in-back-6087802.php.
16. Amina Elahi, "Airbnb to Begin Collecting Chicago Hotel Tax Feb. 15," *Chicago Tribune,* January 30, 2015, http://www.chicagotribune.com/bluesky/originals/chi-airbnb-chicago-taxes-bsi-20150130-story.html.
17. Emily Badger, "Airbnb Is About to Start Collecting Hotel Taxes in More Major Cities, Including Washington," *Washington Post,* January 29, 2015, https://www.washingtonpost.com/news/wonk/wp/2015/01/29/airbnb-is-about-to-start-collecting-hotel-taxes-in-more-major-cities-including-washington/.
18. Dustin Gardiner, "Airbnb to Charge Sales Tax on Phoenix Rentals," *Arizona Republic,* June 26, 2015, http://www.azcentral.com/story/news/local/phoenix/2015/06/25/airbnb-charge-sales-tax-phoenix-rentals/29283651/.
19. Vince Lattanzio, "You'll No Longer Be Breaking the Law Renting on Airbnb," NBC 10, June 19, 2015, http://www.nbcphiladelphia.com/news/local/Youll-No-Longer-Be-Breaking-the-Law-by-Renting-on-Airbnb-308272641.html.
20. "Amsterdam and Airbnb Sign Agreement on Home Sharing and Tourist Tax," *I Amsterdam,* December 18, 2014, http://www.iamsterdam.com/en/media-centre/city-hall/press-releases/2014-press-room/amsterdam-airbnb-agreement.
21. "A Major Step Forward in Paris and France—Une Avancée Majeure En France," Airbnb Action, March 26, 2014, https://www.airbnbaction.com/major-step-forward-paris-france/.
22. Sean O'Neill, "American Hotel Association to Fight Airbnb and Short-Term Rentals," *Tnooz,* April 30, 2014, https://www.tnooz.com/article/american-hotel-association-launches-fightback-airbnb-short-term-rentals/.
23. Josh Dawsey, "Union Financed Fight to Block Airbnb in New York City," *Wall Street Journal,* May 9, 2016, http://www.wsj.com/articles/union-financed-fight-to-block-airbnb-in-new-york-city-1462842763.
24. Jessica Pressler, "The Dumbest Person in Your Building Is Passing Out Keys to Your Front Door!," NYMag.com, September 23, 2014, http://nymag.com/news/features/airbnb-in-new-york-debate-2014-9/.
25. "A.G. Schneiderman Releases Report Documenting Widespread Illegality Across Airbnb's NYC Listings; Site Dominated by Commercial Users," New York State Attorney General, October 1, 2014, http://www.ag.ny.gov/press

-release/ag-schneiderman-releases-report-documenting-widespread-illegality-across-airbnbs-nyc.

26. Carolyn Said, "S.F. Airbnb Law Off to Slow Start; Hosts Say It's Cumbersome," *SFGate,* March 3, 2015, http://www.sfgate.com/business/article/S-F-Airbnb-law-off-to-slow-start-hosts-say-6110902.php.
27. Philip Matier and Andrew Ross, "Airbnb Backers Invest Big on Chiu's Campaign Against Campos," *SFGate,* October 15, 2014, http://www.sfgate.com/bayarea/article/Airbnb-backers-invest-big-on-Chiu-s-campaign-5822784.php.
28. Dara Kerr, "San Francisco Mayor Signs Landmark Law Making Airbnb Legal," CNET.com, October 28, 2014, http://www.cnet.com/news/san-francisco-mayor-makes-airbnb-law-official/.
29. "Historic Day for Home Sharing in San Francisco," Airbnb Action, October 27, 2014, https://www.airbnbaction.com/historic-day-home-sharing-san-francisco/.
30. Peter Shih, http://susie-c.tumblr.com/post/58375244538/peter-shih-wrote-this-yesterday-when-everyone.
31. Carolyn Said, "Would SF Prop. F Spur Airbnb Suits, with Neighbor Suing Neighbor?," *SFGate,* August 31, 2015, http://www.sfgate.com/business/article/Would-SF-Prop-F-spur-Airbnb-suits-with-neighbor-6472468.php.
32. Daniel Hirsch, "Report: Airbnb Cuts into Housing, Should Share Data," MissionLocal, May 14, 2015, http://missionlocal.org/2015/05/report-airbnb-cuts-into-housing-should-give-up-data/.
33. Booth Kwan, "Protesters Occupy Airbnb HQ Ahead of Housing Affordability Vote," *Guardian,* November 2, 2015, https://www.theguardian.com/us-news/2015/nov/02/airbnb-san-francisco-headquarters-occupied-housing-protesters.
34. Eric Johnson, "'Re/Code Decode': Airbnb CEO Brian Chesky Talks Paris Terror Attacks, San Francisco Politics," *Recode,* November 30, 2015, http://www.recode.net/2015/11/30/11621000/recode-decode-airbnb-ceo-brian-chesky-talks-paris-terror-attacks-san.
35. Carolyn Said, "Prop. F: S.F. Voters Reject Measure to Restrict Airbnb Rentals," *SFGate,* November 4, 2015, http://www.sfgate.com/bayarea/article/Prop-F-Measure-to-restrict-Airbnb-rentals-6609176.php.
36. "Berlin Authorities Crack Down on Airbnb Rental Boom," *Guardian,* May 1, 2016, https://www.theguardian.com/technology/2016/may/01/berlin-authorities-taking-stand-against-airbnb-rental-boom.
37. Yuji Nakamura, "Airbnb Faces Major Threat in Japan, Its Fastest-Growing Market," Bloomberg.com, February 18, 2016, http://www.bloomberg.com/news/articles/2016-02-18/fastest-growing-airbnb-market-under-threat-as-japan-cracks-down.

38. Murray Cox and Tom Slee, "How Airbnb's Data Hid the Facts in New York City," InsideAirbnb.com, February 10, 2016, http://insideairbnb.com/reports/how-airbnbs-data-hid-the-facts-in-new-york-city.pdf.
39. Erik Larson and Andrew M. Harris, "Airbnb Sued, Accused of Ignoring Hosts' Race Discrimination," Bloomberg.com, May 18, 2016, http://www.bloomberg.com/news/articles/2016-05-18/airbnb-sued-over-host-s-alleged-discrimination-against-black-man.
40. Benjamin Edelman, "Preventing Discrimination at Airbnb," BenEdelman.org, June 23, 2016, http://www.benedelman.org/news/062316-1.html.
41. Kristen Clarke, "Does Airbnb Enable Racism?," *New York Times,* August 23, 2016, http://www.nytimes.com/2016/08/23/opinion/how-airbnb-can-fight-racial-discrimination.html.
42. Melissa Mittelman, "Airbnb Hires Eric Holder to Develop Anti-Discrimination Plan," Bloomberg.com, July 20, 2016, http://www.bloomberg.com/news/articles/2016-07-20/airbnb-hires-eric-holder-to-develop-anti-discrimination-plan.
43. "Airbnb CEO on Discrimination: 'I Think We Were Late to This Issue,'" *Fortune,* July 13, 2016, http://fortune.com/2016/07/13/airbnb-chesky-discrimination/.
44. Max Chafkin and Eric Newcomer, "Airbnb Faces Growing Pains as It Passes 100 Million Guests," July 11, 2016, Bloomberg.com, http://www.bloomberg.com/news/articles/2016-07-11/airbnb-faces-growing-pains-as-it-passes-100-million-users.
45. Dennis Schaal, "Expedia Buys HomeAway for $3.9 Billion," *Skift,* November 4, 2015, https://skift.com/2015/11/04/expedia-acquires-homeaway-for-3-9-billion/.
46. Amy Plitt, "NYC Hotel Rates May Be Dropping Thanks to Airbnb," *Curbed NY,* April 19, 2016, http://ny.curbed.com/2016/4/19/11458984/airbnb-new-york-hotel-rates-dropping.

Chapter 12: Global Mega-Unicorn Death Match!

1. Rhiannon Williams and Matt Warman, "London at a Standstill but Uber Claims Taxi Strike Victory," *Telegraph,* June 11, 2014, http://www.telegraph.co.uk/technology/news/10892224/London-at-a-standstill-but-Uber-claims-taxi-strike-victory.html.
2. James Titcomb, "What Is Uber and Why Does TFL Want to Crack Down on It?," *Telegraph,* September 30, 2015, http://www.telegraph.co.uk/technology/uber/11902093/What-is-Uber-and-why-does-TfL-want-to-crack-down-on-it.html.

3. Oscar Williams-Grut, "Taxi Drivers Caused Chaos at London's City Hall after Boris Johnson Called Them 'Luddites,'" *Business Insider,* September 16, 2015, http://www.businessinsider.com/london-mayor-boris-johnsons-question-time-disrupted-by-uber-protest-2015-9.
4. James Titcomb, "Uber Wins Victory in London as TFL Drops Proposals to Crack Down on App," *Telegraph,* January 20, 2016, http://www.telegraph.co.uk/technology/uber/12109810/Uber-wins-victory-in-London-as-TfL-drops-proposals-to-crack-down-on-app.html.
5. Sam Schechner, "Uber Meets Its Match in France," *Wall Street Journal,* September 18, 2015, http://www.wsj.com/articles/uber-meets-its-match-in-france-1442592333.
6. "Perquisitions Au Siège D'Uber France," LeMonde.fr, March 17, 2015, http://www.lemonde.fr/societe/article/2015/03/17/perquisitions-au-siege-d-uber-france_4595591_3224.html.
7. Romain Dillet, "Uber France Leaders Arrested for Running Illegal Taxi Company," *TechCrunch,* June 29, 2015, https://techcrunch.com/2015/06/29/uber-france-leaders-arrested-for-running-illegal-taxi-company/.
8. Anne-Sylvaine Chassany and Leslie Hook, "Uber Found Guilty of Starting 'Illegal' Car Service by French Court," *Financial Times,* http://www.ft.com/cms/s/0/3d65be7a-2e22-11e6-bf8d-26294ad519fc.html.
9. Philip Willan, "Italian Court Bans UberPop, Threatens Fine," *PCWorld,* May 26, 2015, http://www.pcworld.com/article/2926752/italian-court-bans-uberpop-threatens-fine.html.
10. "Why UberPop Is Being Scrapped in Sweden," *Local SE,* May 11, 2016, http://www.thelocal.se/20160511/heres-why-uberpop-is-being-scrapped-in-sweden.
11. Lisa Fleisher, "Uber Shuts Down in Spain After Telcos Block Access to App," *Wall Street Journal,* December 21, 2014, http://blogs.wsj.com/digits/2014/12/31/uber-shuts-down-in-spain-after-telcos-block-access-to-its-app/; Maria Vega Paul, "Uber Returns to Spanish Streets in Search of Regulatory U-Turn," Reuters, March 30, 2016, http://www.reuters.com/article/us-spain-uber-tech-idUSKCN0WW0AO.
12. Mark Scott, "Uber's No-Holds-Barred Expansion Strategy Fizzles in Germany," *New York Times,* January 3, 2016, http://www.nytimes.com/2016/01/04/technology/ubers-no-holds-barred-expansion-strategy-fizzles-in-germany.html?_r=0.
13. Brad Stone and Lulu Yilun Chen, "Uber Slayer: How China's Didi Beat the Ride-Hailing Superpower," *Bloomberg Businessweek,* October 6, 2016, https://www.bloomberg.com/features/2016-didi-cheng-wei/.

14. "Hangzhou Kuaizhi Technology (Kuaidi Dache) closes venture funding," Financial Deals Tracker, *MarketLine,* April 10, 2013.
15. Zheng Wu and Vanessa Piao, "Didi Dache, a Chinese Ride-Hailing App, Raises $700 Million," *New York Times,* December 10, 2014, http://dealbook.nytimes.com/2014/12/10/didi-dache-a-chinese-ride-hailing-app-raises-700-million/.
16. "Baidu to Buy Uber Stake in Challenge to Alibaba in China," Bloomberg.com, December 17, 2014, http://www.bloomberg.com/news/articles/2014-12-17/baidu-to-buy-uber-stake-in-challenge-to-alibaba-for-car-booking.
17. Rose Yu, "For Cabs in China, Traffic Isn't Only Woe," *Wall Street Journal,* January 14, 2015, http://www.wsj.com/articles/china-taxi-drivers-continue-striking-over-growing-ride-hailing-services-1421239127.
18. Gillian Wong, "Uber Office Raided in Southern Chinese City," *Wall Street Journal,* May 1, 2015, http://www.wsj.com/articles/uber-office-raided-in-southern-chinese-city-1430483542.
19. Charles Clover, "Uber in Taxi War of Attrition with Chinese Rival Didi Dache," *Financial Times,* http://www.ft.com/cms/s/0/7de53f7a-5088-11e5-b029-b9d50a74fd14.html#axzz4FHeEnQUa.
20. Ibid.
21. Ibid.
22. Tatiana Schlossberg, "New York City Council Discusses Cap on Prices Charged by Car-Service Apps During Peak Times," *New York Times,* January 12, 2015, http://www.nytimes.com/2015/01/13/nyregion/new-york-city-council-discusses-cap-on-prices-charged-by-car-service-apps-during-peak-times.html.
23. Annie Karni, "Uber Loses TLC Appeal to Turn over Trip Data," *New York Daily News,* January 22, 2015, http://www.nydailynews.com/news/politics/uber-loses-tlc-deal-turn-trip-data-article-1.2087718.
24. Michael M. Grynbaum, "Taxi Industry Opens Wallet for De Blasio, a Chief Ally," *New York Times,* July 17, 2012, http://www.nytimes.com/2012/07/18/nyregion/de-blasio-reaps-big-donations-from-taxi-industry-he-aided.html.
25. Tim Fernholz, "The Latest Round in Uber's Battle for New York City, Explained," *Quartz,* June 30, 2015, http://qz.com/441608/the-latest-round-in-ubers-battle-for-new-york-city-explained/.
26. Andrew J. Hawkins, "City Yields to Uber on App Rules," *Crain's New York Business,* June 18, 2015, http://www.crainsnewyork.com/article/20150618/BLOGS04/150619866/city-yields-to-uber-on-app-rules.

27. Colleen Wright, "Uber Says Proposed Freeze on Licenses in New York City Would Limit Competition," *New York Times,* June 30, 2015, http://www.nytimes.com/2015/07/01/nyregion/uber-says-proposed-freeze-on-licenses-would-limit-competition.html.
28. Kirstan Conley and Carl Campanile, "Cuomo Drops Bombshell on De Blasio over Uber," *New York Post,* July 22, 2015, http://nypost.com/2015/07/22/cuomo-drops-bombshell-on-de-blasio-over-uber/.
29. Dan Rivoli, "De Blasio's Multimillion-Dollar Study Blames Deliveries, Construction and Tourism for Traffic Congestion—Not Uber," *New York Daily News,* January 15, 2016, http://www.nydailynews.com/new-york/de-blasio-study-blames-construction-tourism-traffic-article-1.2498253.
30. Ryan Parry, "Exclusive: Luxury Hotels, All-Night Partying at Posh Clubs, Endless Freebies," *Daily Mail Online,* October 1, 2015, http://www.dailymail.co.uk/news/article-3256259/Luxury-hotels-night-partying-posh-clubs-endless-freebies-Uber-hosts-SECRET-Sin-City-team-building-junket-4-800-employees-world-no-drivers-please.html.
31. "An Uber Impact: 20,000 Jobs Created on the Uber Platform Every Month," Uber, May 27, 2014, https://newsroom.uber.com/an-uber-impact-20000-jobs-created-on-the-uber-platform-every-month-2/.
32. Justin Singer, "Beautiful Illusions: The Economics of UberX," *Valleywag,* June 11, 2014, http://valleywag.gawker.com/beautiful-illusions-the-economics-of-uberx-1589509520; Felix Salmon, "How Well UberX Pays, Part 2," *Medium,* June 8, 2014, https://medium.com/@felixsalmon/how-well-uberx-pays-part-2-cbc948eaeeaf#.wc3njxtdz.
33. Alex Barinka, Eric Newcomer, and Lulu Chen, "Uber Backers Said to Push for Didi Truce in Costly China War," Bloomberg.com, July 20, 2016, https://www.bloomberg.com/news/articles/2016-07-20/uber-investors-said-to-push-for-didi-truce-in-costly-china-fight.

Epilogue

1. Max Chafkin, "Uber's First Self-Driving Fleet Arrives in Pittsburgh This Month," Bloomberg.com, August 18, 2016, http://www.bloomberg.com/news/features/2016-08-18/uber-s-first-self-driving-fleet-arrives-in-pittsburgh-this-month-is06r7on.

Index

Index

Index

About the Author

Brad Stone is senior executive editor of global technology at Bloomberg News. He is the author of the *New York Times* bestseller *The Everything Store: Jeff Bezos and the Age of Amazon,* winner of the 2013 Financial Times and Goldman Sachs Business Book of the Year Award. He has covered Silicon Valley for nearly twenty years, writing for *Bloomberg Businessweek,* the *New York Times,* and *Newsweek.* He lives in San Francisco.